I0014143

Fikret Dülger, Edgar Sánchez-Sinencio

Integrated RF Building Blocks for Wireless Communication Transceivers

Fikret Dülger, Edgar Sánchez-Sinencio

Integrated RF Building Blocks for Wireless Communication Transceivers

The Design of LC VCOs, Prescalers, and Q-enhanced LC Bandpass Filters on Silicon

VDM Verlag Dr. Müller

Impressum/Imprint (nur für Deutschland/ only for Germany)
Bibliografische Information der Deutschen Nationalbibliothek: Die Deutsche Nationalbibliothek verzeichnet diese Publikation in der Deutschen Nationalbibliografie; detaillierte bibliografische Daten sind im Internet über http://dnb.d-nb.de abrufbar.

Coverbild: www.purestockx.com

Verlag: VDM Verlag Dr. Müller Aktiengesellschaft & Co. KG
Dudweiler Landstr. 125 a, 66123 Saarbrücken, Deutschland
Telefon +49 681 9100-698, Telefax +49 681 9100-988, Email: info@vdm-verlag.de
Zugl.: College Station, Texas A&M University, Diss.,2002

Herstellung in Deutschland:
Schaltungsdienst Lange o.H.G., Zehrensdorfer Str. 11, D-12277 Berlin
Books on Demand GmbH, Gutenbergring 53, D-22848 Norderstedt
Reha GmbH, Dudweiler Landstr. 99, D- 66123 Saarbrücken
ISBN: 978-3-639-07430-7

Imprint (only for USA, GB)
Bibliographic information published by the Deutsche Nationalbibliothek: The Deutsche Nationalbibliothek lists this publication in the Deutsche Nationalbibliografie; detailed bibliographic data are available in the Internet at http://dnb.d-nb.de.

Cover image: www.purestockx.com

Publisher:
VDM Verlag Dr. Müller Aktiengesellschaft & Co. KG
Dudweiler Landstr. 125 a, 66123 Saarbrücken, Germany
Phone +49 681 9100-698, Fax +49 681 9100-988, Email: info@vdm-verlag.de

Produced in USA and UK by:
Lightning Source Inc., 1246 Heil Quaker Blvd., La Vergne, TN 37086, USA
Lightning Source UK Ltd., Chapter House, Pitfield, Kiln Farm, Milton Keynes, MK11 3LW, GB
BookSurge, 7290 B. Investment Drive, North Charleston, SC 29418, USA
ISBN: 978-3-639-07430-7

To Jimena and Santiago Deniz ..

PREFACE

This book is a slightly modified and updated version of a Ph.D. dissertation defended in 2002 at Texas A&M University, [1]. It includes discussions on the design of fully-integrated Voltage Controlled Oscillators (VCOs), prescalers and Q-enhancement LC bandpass filters on silicon within the framework of Radio Frequency (RF) integrated circuit design for wireless communication transceivers.

Chapter I starts with a brief summary of the widely used receiver and transmitter architectures in wireless communication transceivers. The problem definitions and the motivations behind the proposed research are introduced next, followed by a short literature survey. The chapter is concluded with the specifics of the proposed research.

Fully integrated negative conductance VCO design examples at GHz frequencies in Si BiCMOS and CMOS technologies are provided in Chapter II, [2]. The advantages and shortcomings of the structures are briefly presented together with the associated design trade-offs. Phase noise of a CMOS VCO is calculated based on the Linear Time-Variant Impulse Sensitivity Function (ISF) theory and compared to the simulations. Measurement results of several fully integrated CMOS VCOs designed with 0.5μm and 0.35μm CMOS technologies at 2.2GHz are given in Chapter V, where all the integrated circuit measurements are presented.

In Section C of Chapter II, design considerations in a CMOS LC symmetric VCO design with spiral inductors and bulk-tuned PMOS capacitors are presented, [3]. Source-degeneration resistors used to linearize the negative resistance generator provide additional design degrees of freedom, trading the close-in phase noise with the phase noise at large offsets. The impact of the well resistance of the bulk-tuned PMOS capacitor on the tuning and quality factor of the varactor structure was demonstrated.

Measured tuning range around 2.15GHz is 3.5% due to the limitation of the well resistance of the bulk-tuned capacitors. Phase noise of the VCO was calculated based on the Linear Time-Variant Impulse Sensitivity Function (ISF) theory and compared to the measured results. Spiral inductors with Q of 2.5 at 2GHz limit the phase noise to -126dBc/Hz at 3MHz offset from a 2.19GHz carrier. The VCO implemented in a 0.5 μm CMOS technology sinks 4mA from a 3V supply. Calculated, simulated and measured phase noise results are in good agreement. The measurement results are provided in Chapter V.

In Chapter III, after a short introduction regarding the importance of prescaler performance in PLLs, design considerations in a dual modulus divide by 32/33 prescaler designed with a 0.6μm BiCMOS process are presented, [4]. Care was taken to design the ECL-based circuits to operate with as low supply voltage and current consumption as possible. The phase noise contribution of the integrated bandgap bias network is demonstrated through simulations. The trade-off between the power consumption and the phase noise is pointed out and some guidelines are provided to improve the noise performance. Measurements included in Chapter V confirm the functionality of the prescaler with a 2.5V supply drawing around 2.3mA at 2.35 GHz with an input sensitivity between -24dBm and 12dBm. The circuit operates with a supply voltage down to 2.1V but with limited input sensitivity. A short survey on Injection-Locked frequency dividers concludes Chapter III.

The topic of Chapter IV is the design and analysis of a 2.1GHz, 1.3V, 5mW, fully integrated Q-enhancement LC Bandpass Biquad programmable in peak gain, Q and f_o, implemented in a 0.35μm standard CMOS technology, [5], [6]. The filter uses a resonator built with spiral inductors and inversion-mode PMOS capacitors that provide frequency tuning. The Q tuning is through an adjustable negative-conductance generator. Simplified noise and nonlinearity analyses presented demonstrate the design trade-offs involved in the design. Detailed measurement results are provided in Chapter V. Measured frequency tuning range around 2.1GHz is 13%. Spiral inductors with Q_L of 2 at 2.1GHz limit the spurious-free dynamic range (SFDR) at a minimum of 31dB within the frequency tuning range. The filter sinks 4mA from a 1.3V supply providing a Q of 40 at 2.19GHz with a 1dB compression point dynamic range of 35dB. The circuit operates with supply voltages ranging from 1.2V to 3V. The silicon area is 0.1mm^2. Simulated and measured 1dB compression point results are in good agreement.

Fikret Dülger

Edgar Sánchez-Sinencio

August 2008

ACKNOWLEDGMENTS

The authors would like to acknowledge the invaluable help of Dr.Abdellatif Bellaouar on the Prescaler Design. They express their gratitude to Dr.José Silva-Martínez for the fruitful discussions on the filter design, his valuable suggestions, comments and remarks. Dr.Sherif Embabi and Dr.Maher Abuzaid deserve special thanks with their support throughout the work. Dr.Uğur Çilingiroğlu is acknowledged for his very helpful comments on the bulk-tuned varactor design. Dr.Ranjit Gharpurey's precious comments and suggestions on the symmetrical VCO design and nonlinearity analyses are greatly appreciated. Dr.Devrim Yılmaz Aksın deserves special thanks for providing his symbolic analysis tool written in MATLAB for nonlinearity analyses. Dr.Costas Georghiades, Dr.Alberto Garcia-Diaz, and Dr.Urs Kreuter are all acknowledged for their valuable suggestions on the presentation of the measured data in the book. The authors benefited greatly from the technical discussions they had with Michel Frechette, Francesco Dantoni, Dr.Mostafa Elmala, Dr.Shouli Yan and Naveen K. Yanduru and their help is well appreciated. They would like to express their gratitude to Texas Instruments for providing the measurement equipment for the characterization of some of the ICs and for the fabrication of the prescaler chip. The support and invaluable assistance of Dr. Aydın İlker Karşılayan on LaTeX related problems encountered throughout the preparation of the manuscript are very much appreciated.

The last people to mention by the first author are the most valuable people in his life. They are his beloved wife, his most precious son, his lovely mother, father and brother. Their love provides him with the strength to overcome the difficulties in life.

TABLE OF CONTENTS

CHAPTER Page

I INTRODUCTION 1

A. Transceiver Architectures 1
B. Problem Definition and Motivations 9
1. VCOs in Phase Locked Loops 10
2. Prescalers in Phase Locked Loops 13
3. Fully-Integrated RF Bandpass Filters 15
C. A Brief Literature Review 16
D. Specifics of the Proposed Research 21
1. Negative-Conductance LC Voltage-Controlled Oscillators 21
2. Dual-Modulus Prescaler Design in BiCMOS 22
3. Fully-Integrated LC Q-Enhancement RF Bandpass Filters in CMOS 24

II NEGATIVE-CONDUCTANCE LC VOLTAGE-CONTROLLED OSCILLATORS 26

A. VCO in a PLL 26
B. Negative Conductance Voltage-Controlled Oscillators . . . 28
1. Bipolar VCO Design 29
2. CMOS VCO Design 31
C. Design Considerations in a Symmetric Linearized CMOS LC VCO with Bulk-Tuned PMOS Capacitors 38
1. CMOS Voltage-Controlled Oscillator Design 39
a. Spiral Inductors 41
b. Bulk-tuned PMOS Capacitors 42
c. Phase Noise and the Source-Degeneration Resistors 50

III DUAL-MODULUS PRESCALER DESIGN IN BICMOS 66

A. Prescalers in PLLs 66
B. Dual-Modulus BiCMOS Prescaler Design 68
C. Injection-Locked Frequency Dividers 75
1. Circuit Implementations 77

IV FULLY-INTEGRATED LC Q ENHANCEMENT RF BANDPASS FILTERS IN CMOS 83

Page

A. Q-enhancement LC filters 84
B. Circuit Design 87
1. Nonlinearity Considerations 93
a. Contribution of the Negative Conductance Generator 93
b. Contribution of the Input G_m Stage 104
c. Contribution of the Varactor 117
d. Nonlinearity Analysis with Direct Calculation of Nonlinear Responses 122
2. Noise Considerations 131
3. Dynamic Range Simulations 149

V INTEGRATED CIRCUIT MEASUREMENTS 153

A. Negative-Conductance LC oscillators in CMOS 154
1. Fully-Integrated CMOS LC VCOs with NMOS Cross-Coupled Pairs 154
2. A Symmetric Linearized CMOS LC VCO with Bulk-Tuned PMOS Capacitors 167
B. A Dual-Modulus Prescaler in 0.6 μm BiCMOS 181
C. A Fully-Integrated LC Q-Enhancement RF Bandpass Filter in CMOS 190

VI CONCLUSIONS AND FUTURE WORK 216

REFERENCES 219

APPENDIX A 235

1. Nonlinearity of a MOS differential pair 235
2. Nonlinearity of a MOS differential pair with source-degeneration 244

LIST OF TABLES

TABLE Page

I Comparison between the receiver architectures. 6

II Calculated phase noise percentage contributions of the VCO components 1. 59

III Calculated phase noise percentage contributions of the VCO components 2. 60

IV Simulated power consumption percentage contributions of the dual-modulus prescaler building blocks. 72

V Simulated vs. calculated Input 1 dB compression point for various filter quality factors with the input G_m stage as the only nonlinearity contributor. 116

VI Simulated vs. calculated noise related characteristics of the filter (Q=8). 141

VII Simulated vs. calculated noise related characteristics of the filter (Q=20). 145

VIII Calculated normalized Figure-of-Merit (FOM) for some CMOS and Bipolar VCOs in the literature. 174

IX Calculated dynamic range figures based on the measurements with $V_{dd} = 1.3$V. 205

X Performance comparison of CMOS, BiCMOS and Bipolar integrated RF filters in the literature. 212

XI Comparison between the specifications of an image reject filter in a multi standard receiver for DCS/UMTS and the measured performance of the RF filter. 215

LIST OF FIGURES

FIGURE		Page
1	A wireless communication transceiver front-end.	1
2	A Dual-IF heterodyne receiver.	2
3	A homodyne receiver with quadrature downconversion.	3
4	Hartley image reject receiver.	4
5	Weaver image reject receiver.	5
6	A digital-IF receiver.	5
7	A direct conversion transmitter.	7
8	A two-step transmitter.	7
9	Phase-locked loop: (a) basic block diagram, (b) phase noise of a first-order loop with a low noise input.	11
10	Impact of the prescaler phase noise on the phase noise of a first-order loop with a low noise input.	15
11	Negative conductance voltage-controlled oscillator: (a) concept, (b) circuit implementation in CMOS.	22
12	Block diagram of the dual-modulus divide by 32/33 prescaler.	23
13	Phase-locked loop: (a) basic block diagram, (b) phase noise of a first-order loop with noiseless input.	27
14	Negative conductance voltage-controlled oscillator: (a) concept, (b) circuit implementation in CMOS.	29
15	Bipolar voltage-controlled oscillator.	30
16	CMOS voltage-controlled oscillator.	32

FIGURE Page

17 SpectreRF simulated vs. analytically calculated phase noise of the CMOS VCO. 36

18 Simulated phase noise of the NMOS-only VCO @ 3MHz offset vs. the quality factor of the tank inductor. 37

19 CMOS voltage-controlled oscillator: (a) concept, (b) circuit implementation. 40

20 Asymmetrical model of the integrated spiral inductor used in the differential VCO simulations. 42

21 Simulated spiral inductor quality factor vs. frequency. 42

22 PMOS bulk-tuned capacitor as tank varactor. 43

23 Simulated effect of the well resistance of the PMOS bulk-tuned capacitor on the tuning. 44

24 Simulated effect of the well resistance of the PMOS bulk-tuned capacitor on the quality factor. 45

25 Small-signal equivalent circuit of the capacitive component of the bulk-tuned PMOS varactor. 46

26 Comparison between the simulated and the analytically calculated transfer characteristic of the bulk-tuned PMOS capacitor. 50

27 CMOS voltage-controlled oscillator with noise sources for phase noise analysis. 54

28 Simulated ISFs of the cross-coupled transistors and the tail transistor in the CMOS VCO. 57

29 Simulated ISFs of the source-degeneration resistors and the diode-connected current source transistor in the CMOS VCO. 58

30 Calculated (with Eqs. (2.17) and (2.21)) vs. SpectreRF simulated phase noise of the CMOS VCO. 59

31 Simulated phase noise of the CMOS VCO @ 3MHz offset vs. the quality factor of the tank inductor. 61

Page

32 Simulated effect of the source degeneration resistors on the close-in phase noise of the CMOS VCO. 62

33 Simulated effect of the source degeneration resistors on the phase noise at large offsets from the carrier of the CMOS VCO. 63

34 Simulated effect of the source degeneration resistors on the oscillation amplitude of the CMOS VCO. 64

35 Phase-locked loop: (a) basic block diagram, (b) impact of the divider phase noise on the phase noise of a first-order loop with a low noise input. 67

36 Block diagram of the dual-modulus prescaler. 69

37 Bandgap bias generator of the prescaler. 70

38 The input buffer. 71

39 The ECL-based master and slave DFF. 72

40 The true BiCMOS NOR gate with ECL and CMOS level inputs. . . 73

41 Simulated phase noise of the prescaler. 74

42 Model for an injection-locked frequency divider. 76

43 Single-ended injection-locked frequency divider. 77

44 Differential injection-locked frequency divider (DILFD). 79

45 Shunt-peaking in a differential injection-locked frequency divider for locking range enhancement. 81

46 Q-enhancement LC bandpass biquad: (a) concept, (b) circuit implementation in CMOS. 85

47 Schematic of the CMOS bandpass filter. 87

48 Asymmetrical model of the integrated spiral inductor used in the CMOS bandpass filter simulations. 89

FIGURE Page

49 PMOS capacitor in inversion-mode as the frequency tuning element. 90

50 Simulated capacitance-voltage transfer characteristic and the quality factor of the PMOS capacitor in inversion-mode. 91

51 Simulated 1dB compression point dynamic range versus the quality factor of the inductor with the filter Q=40 and f_o=2.28GHz. . . . 92

52 Cross-coupled differential pair as negative conductance generator. . . 94

53 Two equivalent small signal circuits of the cross-coupled differential pair. 95

54 Cross-coupled differential pair for analysis of the large signal negative conductance. 96

55 Differential conductance transfer characteristic of the nonlinear negative conductance generator with $I_{SS} = 4mA$ and $V_{GS} - V_T = 385mV$. 98

56 Differential conductance vs. the differential voltage of the nonlinear negative conductance generator with $I_{SS} = 4mA$ and $V_{GS} - V_T = 385mV$. 99

57 Comparison of the differential current - differential voltage transfer characteristics using the simulated g_m value in lieu of the $2I_D/(V_{GS} - V_T)$ in the analytical expression, with $I_{SS} = 4mA$ and $V_{GS} - V_T = 385mV$. 101

58 Comparison of the differential conductance transfer characteristics using the simulated g_m value in lieu of the $2I_D/(V_{GS} - V_T)$ in the analytical expression with $I_{SS} = 4mA$ and $V_{GS} - V_T = 385mV$. . . . 102

59 Equivalent circuit of the filter for nonlinearity contribution analysis of the negative conductance generator. 102

60 Differential pair for the analysis of the large signal transconductance. 104

61 Linear gain characteristic of the filter with only the nonlinearity of the input G_m stage included at $f_{in} = 2.3GHz$ and $Q = 40$. 110

62 Gain compression of the filter with only the nonlinearity of the input G_m stage included at $f_{in} = 2.3GHz$ and $Q = 40$. 112

Page

63 Fundamental output power versus the input power of the filter with only the nonlinearity of the input G_m stage included at $f_{in} = 2.3GHz$ and $Q = 40$. 113

64 Third harmonic at the output of the filter with only the nonlinearity of the input G_m stage included at $f_{in} = 2.3GHz$ and $Q = 40$. . 114

65 Capacitance-voltage transfer characteristic of the varactor. 119

66 Equivalent circuit of the filter for nonlinearity contribution analysis of the varactor. 120

67 Frequency characteristic of the magnitude of the small-signal voltage gain of the filter. 121

68 Simulated and theoretical 1dB compression points of the filter with the negative conductance generator as the only nonlinearity contributor, versus the filter Q. 130

69 Simulated and theoretical 1dB compression points of the filter with the varactor as the only nonlinearity contributor versus the filter Q. 130

70 Cross-coupled differential pair as negative conductance generator and its small-signal equivalent circuit. 132

71 Single-ended equivalent circuit of the filter for noise analysis. 133

72 Frequency characteristic of the magnitude of the filter's voltage gain with Q=8. 135

73 Output noise power spectral density of the filter with Q=8. 138

74 Noise figure of the filter with Q=8. 140

75 Frequency characteristic of the magnitude of the voltage gain of the filter with Q=20. 142

76 Output noise power spectral density of the filter with Q=20. 143

77 Noise figure of the filter with Q=20. 144

FIGURE Page

78 Effect of the gate resistances on the noise figure of the filter with Q=20.147

79 Simulated 1dB compression point and spurious-free dynamic range versus the quality factor of the filter. 151

80 Test setup in the laboratory used for the characterization of the ICs. 153

81 Test setup in the laboratory used for the on-wafer characterization of the ICs. 154

82 Die micrograph (0.28mm X 0.33mm) of the VCO with HP 0.5μm CMOS. 155

83 PCB designed to test packaged VCOs with HP 0.5μm CMOS. 155

84 PCB designed for RF characterization of the packaged VCOs with HP 0.5μm CMOS. 156

85 Schematic representation of the filtering on the bias and supply lines on the PCBs. 156

86 Distribution of the oscillation frequency among the fabricated NMOS VCOs for $V_{bulk} = 2.5V$ with HP 0.5μm CMOS technology. 158

87 Distribution of the single-ended fundamental power among the fabricated NMOS VCOs with HP 0.5μm CMOS technology. 159

88 Frequency spectrum of the HP 0.5μm NMOS VCO. 160

89 Frequency spectrum of the harmonic contents of the HP 0.5μm NMOS VCO. 161

90 Frequency tuning characteristic of the HP 0.5μm NMOS VCO. . . . 162

91 Fundamental output power over the frequency tuning of the HP 0.5μm NMOS VCO. 162

92 Phase noise of the VCO with HP 0.5μm CMOS at $f_{osc} = 2.14GHz$. . 163

93 Phase noise of the VCO with HP 0.5μm CMOS at $f_{osc} = 2.2GHz$. . . 163

Page

94 Die micrograph (0.26mm X 0.4mm) of the TSMC 0.35μm CMOS VCO.164

95 Frequency spectrum of the TSMC 0.35μm CMOS VCO with Vdd=1.3V.164

96 Harmonic contents at the output of the TSMC 0.35μm CMOS VCO with Vdd=1.3V. 165

97 Measured tuning range of the TSMC 0.35μm CMOS VCO with $V_{dd} = 1.8V$. 165

98 Phase noise of the VCO with TSMC 0.35μm CMOS at $f_{osc} = 2.15GHz$.166

99 Die micrograph (0.3mm X 0.4mm) of the CMOS voltage-controlled oscillator including the output buffer. 167

100 Distribution of the oscillation frequency among the fabricated CMOS VCOs for $V_{bulk} = 2.5V$ with HP 0.5μm CMOS technology. 168

101 Distribution of the single-ended fundamental power among the fabricated CMOS VCOs with HP 0.5μm CMOS technology. 168

102 Measured frequency spectrum of the CMOS VCO. 169

103 Measured harmonic content of the single-ended output of the CMOS VCO. 170

104 Measured harmonic content of the differential output of the CMOS VCO. 171

105 Measured phase noise of the CMOS VCO at $f_{osc} = 2.19GHz$. 172

106 Measured vs. simulated and calculated (with Eqs. (2.17) and (2.21)) phase noise of the CMOS VCO. 173

107 Measured tuning range of the CMOS bulk-tuned VCO. 175

108 Variation of the single-ended fundamental output power over the tuning range of the CMOS bulk-tuned VCO. 176

109 Variation of the measured phase noise at 3MHz offset throughout the tuning range. 177

FIGURE Page

110 Comparison between the simulated and measured phase noise with $V_{bulk} = 1.5V$. 177

111 Comparison between the simulated and measured phase noise with $V_{bulk} = 2V$. 178

112 Die micrograph of the chip including the three CMOS voltage-controlled oscillators. 178

113 Measured effect of the source degeneration resistors on the close-in phase noise of the CMOS VCOs. 179

114 Measured effect of the source-degeneration resistors on the phase noise at large offsets from the carrier of the CMOS VCOs. 179

115 Comparison of the measured tuning ranges of the VCOs. 180

116 Variation of the measured phase noise at 1MHz offset throughout the tuning range. 180

117 Die micrograph (0.38mm X 0.35mm) of the prescaler including the bias network and the output buffer. 181

118 PCB designed for the characterization of the prescalers. 182

119 Distribution of the maximum frequency of operation among the fabricated prescalers. 183

120 Frequency spectrum of the prescaler with V_{cc}=2.5V. 183

121 Output of the prescaler with f_{in} = 2.35GHz. 184

122 Measured input sensitivity over the frequency. 185

123 Measured input sensitivity over the supply voltage. 186

124 Measured vs. simulated phase noise of the prescaler in divide-by-33 mode with f_{in} = 2.35GHz. 187

125 Phase noise of the prescaler in divide-by-33 mode with f_{in} = 2.35GHz. 187

Page

126 Phase noise of the prescaler averaged on ten consecutive measurements in divide-by-33 mode with $f_{in} = 2.35$GHz. 188

127 Phase noise of the prescaler averaged on eight consecutive measurements in divide-by-32 mode with $f_{in} = 2.35$GHz. 188

128 Phase noise of the prescaler averaged on twelve consecutive measurements in divide-by-32 mode with $f_{in} = 2$GHz. 189

129 Phase noise of the prescaler for three different test settings. 189

130 Die micrograph (0.26mm X 0.4mm) of the CMOS bandpass filter including the output buffer. 190

131 Test setup for the filter measurements. 191

132 PCB designed for the RF characterization of the filter. 191

133 PCB designed for the functionality test of the filter. 192

134 Measured transfer characteristic of the PMOS varactor at 2GHz for V_{dd}=1.8V. 192

135 Equivalent circuit of the PMOS varactor used in the capacitance extraction. 193

136 Gain tuning of the RF filter with Q=40 and f_o=2.12GHz. 197

137 Frequency tuning (f_o=1.93GHz - 2.19GHz) of the RF filter with $Q \approx$100. 198

138 Q tuning (Q=20 - 170) of the RF filter at f_o=2.16GHz. 199

139 Overlaid $|S_{21}|$ of ten different ICs (a) with the same bias settings and (b) after programmed for the same gain, Q=40 and f_o=2.14GHz. 200

140 1dB compression point measurement of the RF filter at f_o=2.19GHz with Q=40. 201

141 Two tone measurement of the RF filter at f_o=2.19GHz with Q=40. . 202

FIGURE Page

142 Measured third-order intermodulation distortion of the RF filter at f_o=2.19GHz with Q=40. 203

143 Spurious-free dynamic range versus the center frequency of the filter for Q=40. 204

144 Measured third-order intercept point versus the center frequency of the filter for Q=40. 205

145 Measured noise figure versus the center frequency of the filter for Q=40.206

146 Measured input-referred noise floor versus the center frequency of the filter for Q=40. 207

147 Dynamic range versus the power consumption of the filter varied through the tail current of the input g_m stage, for f_o=2.17GHz and Q=40. 208

148 Dynamic range versus the supply voltage of the filter for Q=40 and f_o=2.17GHz. 209

149 Dynamic range versus the filter Q with f_o=2.17GHz and V_{dd}=1.5V. . 210

150 Comparison between the measured and simulated 1dB compression point versus the filter quality factor with f_o=2.17GHz and V_{dd}=1.5V. 211

151 Measured noise figure of the standalone output buffer of the filter. . . 213

152 Measured noise figure of the (filter + buffer) with $V_{dd} = 1.3V$ and $Q = 40$. 214

153 Measured 1-dB compression point of the standalone output buffer of the filter (P_{1dB}=2dBm). 214

154 Measured third-order intercept point of the standalone output buffer of the filter (IIP_3=13.5dBm). 215

155 Differential pair with differential excitation at its inputs. 236

156 AC equivalent circuit of the differential pair. 237

Page

157 Linearized equivalent circuit for the calculation of the first-order responses. 238

158 Linearized equivalent circuit for the calculation of the second-order responses. 239

159 Linearized equivalent circuit for the calculation of the third-order responses. 241

160 Source-degenerated differential pair with differential inputs. 244

161 AC equivalent circuit of the differential pair with source-degeneration. 245

162 Linearized equivalent circuit with source-degeneration for the calculation of the first-order responses. 245

163 Linearized equivalent circuit with source-degeneration for the calculation of the second-order responses. 246

164 Linearized equivalent circuit with source-degeneration for the calculation of the third-order responses. 250

CHAPTER I

INTRODUCTION

A. Transceiver Architectures

The wireless communication systems have become an important part of our daily lives. The ever increasing demand towards lower cost makes the task of the wireless communication system and circuit designers more and more challenging. This demand translates into the circuit specifications as the need of designing the circuits with lower power consumption, with smaller die area but without any compromise from the final goal of higher performance, e.g. lower noise for higher signal-to-noise ratio, higher dynamic range, etc.. The circuit design for wireless communication transceiver front-ends has an additional complexity due to the high frequency of operation involved. The block diagram of a wireless communication transceiver front-end is shown in Fig.1, [7].

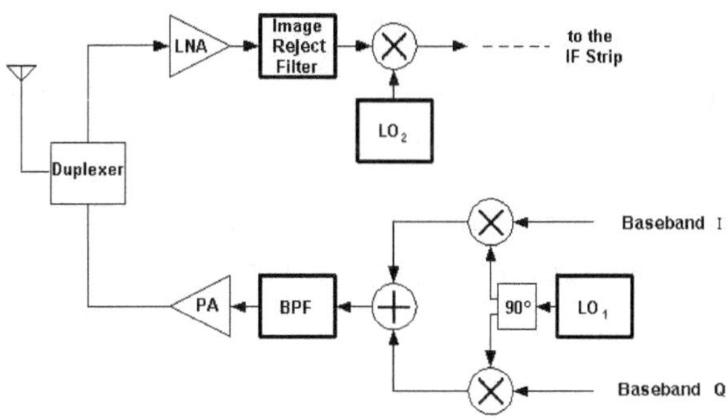

Fig. 1. A wireless communication transceiver front-end.

The journal model is *IEEE Transactions on Automatic Control.*

Front-end is the section of the transceiver next to the antenna where the signal processing is performed at the frequency of the received and transmitted signals. The frequency range of interest in the current wireless communication standards are from around 900MHz to 2.4GHz. The building blocks shown in Fig.1 are the power amplifier (PA), the bandpass filter (BPF), the 90° phase shifter, the quadrature modulator and the first local oscillator (LO_1) in the transmitter and the low noise amplifier (LNA), the image reject filter, the downconversion mixer and the second local oscillator (LO_2) in the receiver. Note that the local oscillators are Phase Locked Loop (PLL) based frequency synthesizers that generate signals with high spectral purity to convert the transmitted and the received information signals in frequency without corrupting them.

Some of the widely used receiver and transmitter architectures are briefly summarized in the following for the sake of providing a rough picture of the framework within which the work in this book fits. The design considerations associated with the architectures are not covered in detail in the following discussion as they are beyond the scope of the research presented in this book. Interested reader may refer to [7] for the advantages and disadvantages of each architecture. The block diagram of a so-called Dual-IF heterodyne receiver is shown in Fig.2, [7].

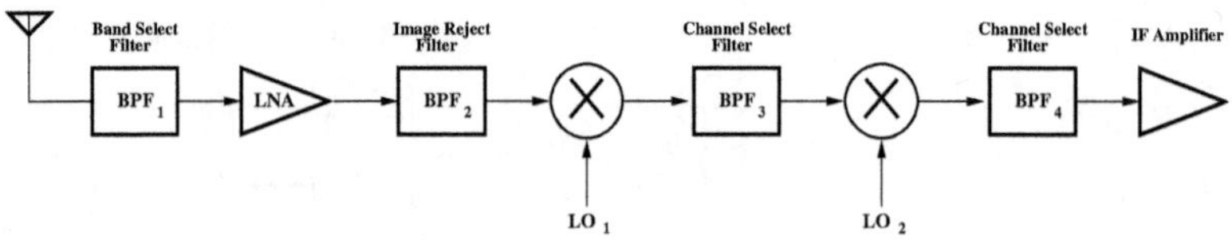

Fig. 2. A Dual-IF heterodyne receiver.

As the name suggests, there are two frequency downconversions in the architecture, each followed by filtering in order to provide the channel selection. The main advantage of multiple downconversions is that the required selectivity of each filter, in other words the Q of each filter is more relaxed with respect to the case of providing the channel selection with one filter. The first downconverter (mixer) is preceeded by an image reject filter in order to prevent the image from corrupting the information signal, [7]. It is important to note that the second downconversion is also susceptible to the image problem and the channel select filter helps alleviate the issue.

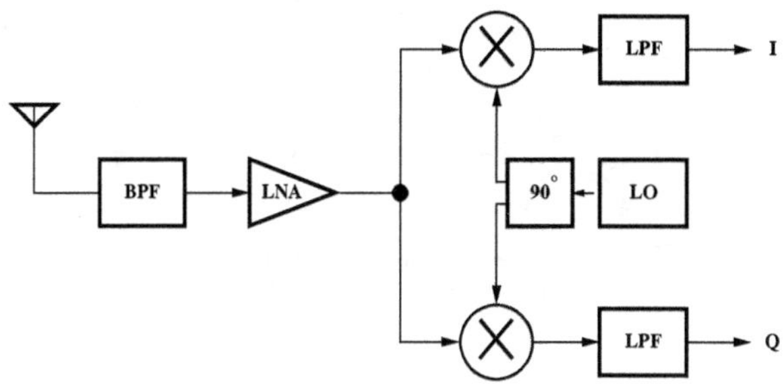

Fig. 3. A homodyne receiver with quadrature downconversion.

One disadvantage of the dual-IF topology is the need for designing two frequency synthesizers as two local oscillator signals denoted as LO_1 and LO_2 are required. It is possible to directly downconvert the RF signal into baseband using the so called Homodyne Receiver shown in Fig.3. The 90° phase shifter following the local oscillator (a PLL based frequency synthesizer in the state-of-the-art solutions) denoted as LO generates the in-phase I and quadrature Q local oscillator signals to downconvert the received signal into the baseband in quadrature signals. The problem of image no longer exists in the architecture as the received signal is downconverted into DC. Namely, no image reject filter is required. On the other hand, downconverting the received information signal directly into DC has some serious disadvantages like DC offsets, 1/f noise etc., [7]. Additionally, the quadrature mixing shown in Fig.3 suffers from the mismatches between the I and Q channel increasing the bit error rate.

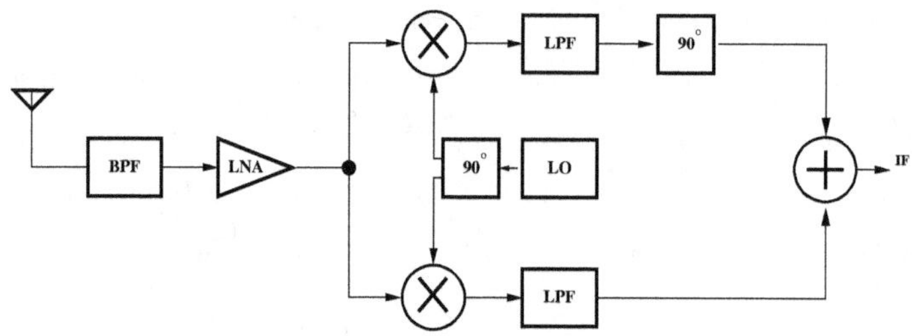

Fig. 4. Hartley image reject receiver.

In order to alleviate the problem of image without having to use image reject filters in the heterodyne receivers, designers sought some other techniques of suppressing the image. The architecture illustrated in Fig.4 mixes the RF signal with the quadrature local oscillator signals and adds an additional 90° phase shift to the downconverted and low-pass filtered in phase I signal before adding the I and Q signals to generate the IF signal to be processed further. Following the signal propagation through the topology yields that the resulting IF signal is the downconverted version of the RF signal not corrupted by the image in the ideal case, [7]. In practice, the amount of image rejection that can be obtained by the structure is limited by the mismatches in the signal paths.

Another architecture that is used to alleviate the image problem is the Weaver architecture shown in Fig.5. Instead of introducing a 90° phase shift at the I branch before the addition as it is done in the Hartley architecture shown in Fig.4, two consecutive quadrature downconversions are performed in the Weaver with lowpass filtering in between. Ideally, the subtraction of the spectrums at the I and Q branches yields the desired signal without any corruption from the image as the signal processing in the architecture produces the same polarities for the desired band and the opposite

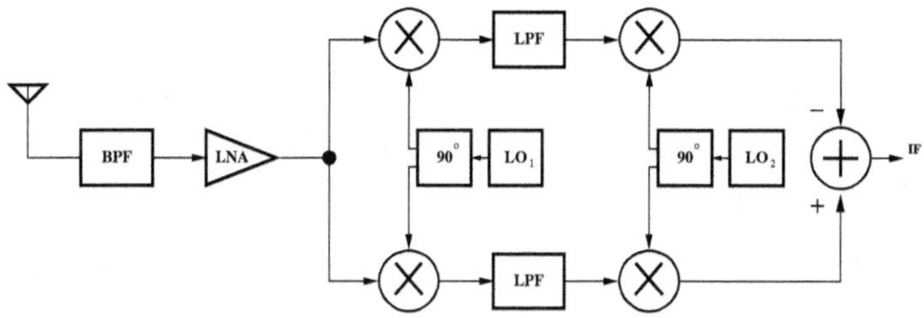

Fig. 5. Weaver image reject receiver.

polarities for the image in the two paths, I and Q, [7]. Weaver architecture also suffers from the mismatches in the quadrature branches. The amount of image rejection is limited by the mismatches in the signal branches as it is the case in the Hartley architecture shown in Fig.4.

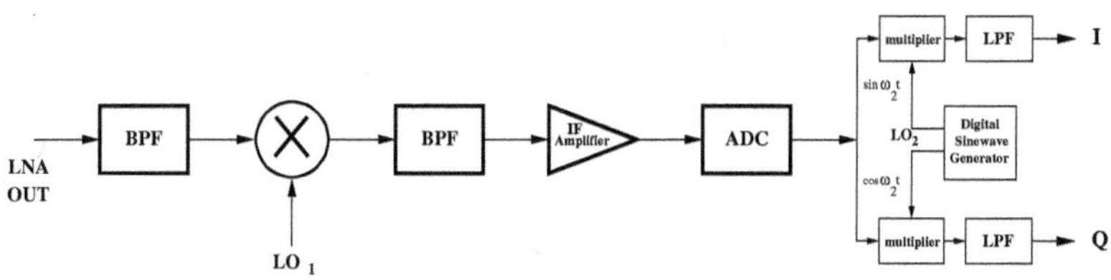

Fig. 6. A digital-IF receiver.

The digital-IF receiver shown in Fig.6 takes advantage of the better efficiency of digital domain operations in the mixing and filtering which are performed in the analog domain in the Dual-IF heterodyne receiver of Fig.2. The main challenge in the digital-IF receiver is the high performance required of the Analog-to-Digital Converter (ADC).

In all the receivers mentioned above, the local oscillator frequency is taken in the vicinity of the RF band. An alternative to that is to sample the RF input at a much lower frequency. This technique, called as "subsampling", makes use of the fact that the RF input signal which is a narrowband signal in the wireless communication environment exhibits a small change from one carrier cycle to another, [7]. The idea behind the subsampling is that a bandpass signal with a bandwidth of Δf can be translated to a lower frequency if sampled with a signal at a frequency of $2\Delta f$ or higher, [7].The simplification of the local oscillator and its associated synthesizer loop due to the much lower frequency required is the main advantage of the subsampling technique. However, a very important drawback is the aliasing of noise due to subsampling. As a result of subsampling by a factor of m, the downconverted noise power of the sampling circuit is amplified by a factor of $2m$, degrading the noise performance substantially, [7].

Table I shows a crude comparison between the receiver architectures briefly covered above. The comparison is not aimed to be a complete one by any means. The parameters involved in the architecture selection are plenty and the priorities in the specific application the receiver is designed for may easily favor an architecture that is seen to be inferior in Table I.

Table I. Comparison between the receiver architectures.

Architecture	Main Challenge	Integration	Complexity
Dual-IF (Heterodyne)	Image rejection	Medium	High
Direct Conversion (Homodyne)	DC offsets, $1/f$ noise	High	Low
Image-Reject (Hartley, Weaver)	Sensitivity to $I-Q$ mismatch	High	Medium
Digital-IF	Design of the ADC	Medium	Medium
Subsampling	Aliasing of noise	High	Low

A direct conversion transmitter where the transmitted frequency from the antenna is equal to the local oscillator signal that is used to upconvert the baseband

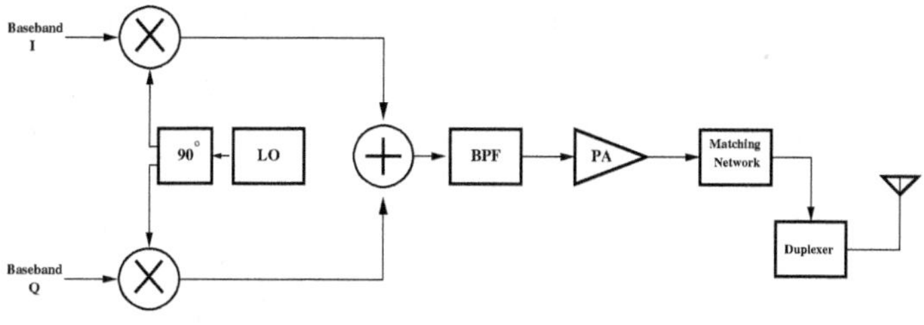

Fig. 7. A direct conversion transmitter.

signal is shown in Fig.7. The transmitted signal amplified by the power amplifier drives the antenna through a matching network and a duplexer that provides different filtering characteristics in the transmit and receive bands in the Frequency-Division Duplexing (FDD) systems. A very important drawback of the direct conversion transmitter is the disturbance of the local oscillator signal by the power amplifier output, through a phenomenon called "injection pulling".

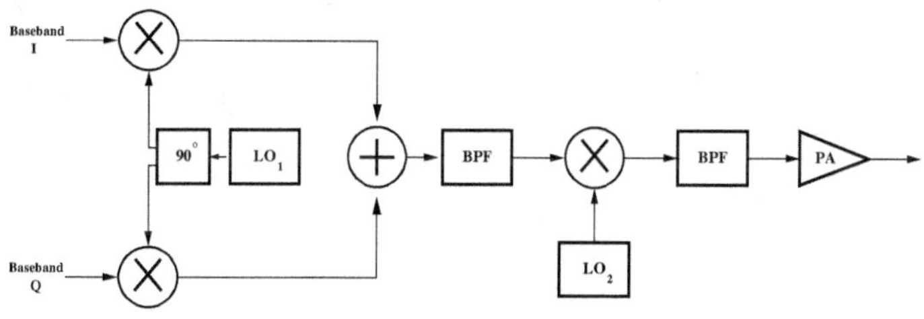

Fig. 8. A two-step transmitter.

A possible approach to circumvent the serious problem of "injection pulling" is to perform the upconversion in two steps as shown in the two-step transmitter architecture shown in Fig.8. Note that, the frequency of the local oscillator signal is far from that of the transmitted signal preventing the disturbance. Another advantage over the direct conversion transmitter is that as the quadrature modulation is performed at lower frequencies, the matching properties are better improving the performance. The main difficulty is the high selectivity required in the second bandpass filter following the second upconversion. As the frequency is high, due to the selectivity requirements, the filter is an expensive passive filter increasing the cost, [7].

B. Problem Definition and Motivations

In the state-of-the-art solutions adopted in the industry today, the bandpass filter in the transmitter and the image reject filter in the receiver are off-chip passive filters that increase the cost and the form factor of the systems designed. The same holds true for some Voltage Controlled Oscillators (VCOs) used in the Local Oscillators. They are either completely off-chip circuits or they use off-chip tank elements, i.e. inductors and varactors, as the resonators with only the active parts integrated. The prescalers in the frequency synthesizers are also very critical blocks to design as they run at the same frequency of the VCO and they are among the major contributors to the power consumption of the synthesizer together with the VCO.

In the proposed research, the undertaken task is to investigate the feasibility of fully-integrated circuit topologies in relatively low cost Silicon CMOS and BiCMOS technologies for the RF bandpass filters, VCOs and prescalers to be used in the wireless communication systems design. The proposed circuit topologies should be demonstrated to work on silicon through integrated circuit measurements. It should be noted that the design of other building blocks such as Low Noise Amplifiers, Mixers and Power Amplifiers are not within the scope of this work. Some recent publications for Low Noise Amplifiers [8]-[14], Mixers [15]-[21] and Power Amplifiers [22]-[28] are listed in the references section at the end of the book for the interested reader.

The frequency synthesizers are mixed-signal systems that accommodate both digital and analog circuitry on the same die. It is important to provide immunity to the sensitive analog sections of the mixed-signal systems from the switching noise generated by the digital sections. The noise that can be coupled through the supply lines and/or the substrate can be treated as a common-mode signal and therefore it is strongly advisable to use fully-differential structures that inherently have some common-mode rejection in the design of the sensitive analog building blocks such as VCOs, amplifiers and filters. Fully-differential topologies are adopted in all the circuits proposed in the research, for this reason.

1. VCOs in Phase Locked Loops

Voltage-Controlled Oscillators (VCOs) are among the key building blocks of wireless communication systems. The quality of the VCO is vital in setting the performance of the Phase Locked Loop (PLL) based frequency synthesizers used in generating the Local Oscillator (LO) signals that convert the transmitted and received information signals in frequency. The trend toward low cost and large scale integration in wireless communications systems design steadily increases the demand for fully integrated VCOs at RF in CMOS and BiCMOS technologies. The main difficulties involved in the design of fully-integrated VCOs for wireless communication systems are the stringent specifications of low noise with low power consumption. In the following paragraph, we would like to emphasize briefly the impact of the VCO phase noise on the overall phase noise of the PLL-based frequency synthesizers, [2].

PLL is a widely used building block in modern communication systems, mostly in synthesizing well-defined, stable local oscillator signals for up and down-converting the transmitted and received signals, respectively, [29]. It is well accepted that the most challenging parts of especially high-frequency PLLs used in wireless communication systems are the VCOs due to the very stringent low noise specifications with low power consumption at GHz frequencies. The increasing demand for complete integration in relatively low cost IC technologies, i.e. CMOS, BiCMOS, without any performance compromise, as the frequency of operation increases steadily complicates the design even further.

Block diagram of a basic PLL is shown in Fig.9(a). A reference signal usually

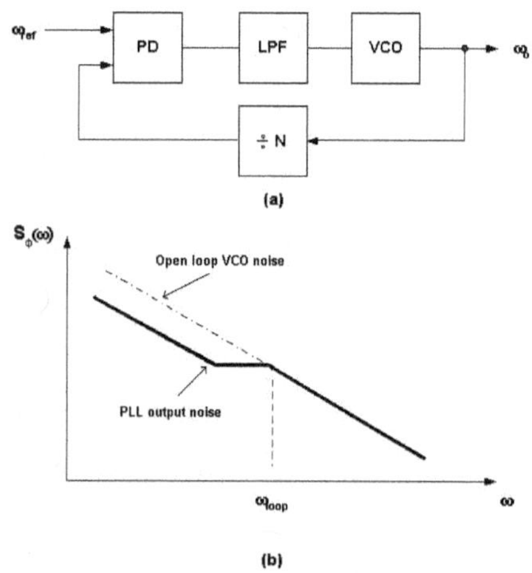

Fig. 9. Phase-locked loop: (a) basic block diagram, (b) phase noise of a first-order loop with a low noise input.

from a low noise crystal oscillator is compared with the divided VCO signal by the Phase Detector (PD) and the resulting signal after low-pass filtered is connected to the VCO as the control voltage, closing the loop. When the loop is in locked condition, the output frequency of the PLL, i.e. the output frequency of the VCO, becomes equal to the reference frequency multiplied with the divider ratio, that is $\omega_o = N \cdot \omega_{ref}$. The demanding phase noise specifications of wireless communication standards sets the major design challenge as achieving low noise at the PLL output. The phase noise of a first-order PLL with a low noise input is depicted in Fig.9(b) for the sake of illustrating the effect of the VCO phase noise on the output noise of the overall PLL system, [30]. Note that, even though the VCO phase noise at offset frequencies lower than the loop bandwidth is suppressed by the loop, the phase noise of the PLL is determined by that of the VCO at offset frequencies larger than the loop bandwidth, denoted as ω_{loop}. On the other hand, the need for synthesizing frequencies with small steps limits the loop bandwidth to small values, [29], thus the VCO phase noise remains unsuppressed at the output of the PLL for some large offset frequencies where the specifications of wireless standards are demanding. This clearly shows the importance of low noise VCO design for a high-performance frequency synthesizer complying with the standards.

The candidates for fully-integrated VCOs to be used in wireless communication systems design are the ring oscillators and the negative conductance LC VCO structures. The design of ring oscillators is well established in the literature and despite their ease of integration on Silicon, the high power consumption they require for low phase noise levels demanded in the wireless communication standards almost rules out their use especially for the Cellular Standards such as GSM 900 MHz and DCS 1800 MHz.

Negative conductance LC differential VCO structures using integrated spiral inductors and varactors as resonators show promising results as more and more efforts are put on improving the quality of the integrated tank passive elements by the designers and process developers. The structure is superior to the ring oscillators with the filtering inherent to the LC tank that serves to reduce the noise level. Consequently, the power consumption levels are also lower compared to those of the ring oscillators. The main challenge is the immaturity of the integrated passive tank elements, i.e. the inductors and the varactors, on Si. The advantages and disadvantages of the negative conductance LC structures from the perspective of complete integration of the VCOs are investigated in the research. Impact of the quality factor of the integrated passive tank elements on the performance are demonstrated analytically and through simulations. Design trade-offs involved are pointed out. Phase noise calculation based on the Linear-Time-Variant Impulse Sensitivity Function theory is performed to enable the understanding of the noise contributions from the devices in the VCO, [30]. Circuit design examples in CMOS technology are provided. Detailed design considerations together with the governing trade-offs and performance metrics based on the simulations of the designs are investigated. The design issues involved in using the bulk terminal of a PMOS transistor to change its effective capacitance at high frequencies are explored. The bulk-tuned structure is used as the tank varactor in a CMOS negative conductance LC VCO. An analytical analysis of the bulk-tuned varactor is provided and the results are supported with simulations. The impact of the integrated tank inductors on the phase noise of the CMOS VCOs through simulations are demonstrated. Several CMOS VCOs are fabricated to validate the theoretical and simulated results through IC measurements. The circuits designed with HP 0.5μm and TSMC 0.35μm CMOS technologies are targeted to operate at around 2GHz.

2. Prescalers in Phase Locked Loops

Prescalers which are basically frequency dividers are among the key building blocks of GHz-range frequency synthesizers used in wireless communication systems, [31], [32]. The quality of the prescaler is also vital in setting the performance of the PLL-based frequency synthesizers used as Local Oscillators (LO), especially from the power consumption point of view. There is a continuing research effort to improve the efficiency (i.e. higher operating frequency with lower power consumption) of the prescalers designed with the mainstream CMOS and BiCMOS technologies, [32], [33]. However, the impact of the prescaler phase noise on the PLL output phase noise has not received much attention in the literature. The noise contributors in the prescalers are hardly investigated in the papers causing a lack of understanding to improve the noise performance of the prescalers designed for wireless communication systems.

The biasing of the integrated circuits (ICs) are provided by bias networks integrated on the same chip. As the ICs are specified to operate within a temperature range, e.g. a range between $-40°C$ and $85°C$ in the industry, the bias networks should provide bias voltages and bias currents that either do not change over the specified temperature range or have a known temperature characteristic so that the circuits are designed accordingly for the performance required. To satisfy this need, bandgap bias networks are used in the ICs. Although bias networks are vital constituents of ICs, no investigation on the impact of the bias network on the phase noise of the prescalers is available in the literature known to the authors. In this research, the demonstration of the phase noise contribution of an integrated bandgap bias network on the phase noise of a dual-modulus prescaler is aimed. The design trade-off between the power consumption and the phase noise is pointed out in a design fabricated with a $0.6\mu m$ BiCMOS technology. The investigation is based on the simulations and the results are validated through integrated circuit measurements. Some design guidelines are also provided to improve the phase noise performance. The target frequency of operation is around 2GHz for the dual-modulus, divide by 32/33 prescaler.

In this section, we would like to emphasize briefly the impact of the prescaler noise on the phase noise performance of the PLL-based frequency synthesizers.

One of the most challenging parts of especially high-frequency PLLs used in wireless communication systems is the prescaler due to the demanding specification of low power consumption and low phase noise at GHz frequencies. Note that in the block diagram shown in Fig.9(a), the prescaler is embedded into the divider in the feedback loop and a fixed-mode prescaler is assumed for the sake of simplicity. The demanding phase noise specifications of wireless communication standards set the major design challenge as achieving low noise at the PLL output. Therefore, the noise performance of the prescalers are to be optimized in designing low noise PLLs.

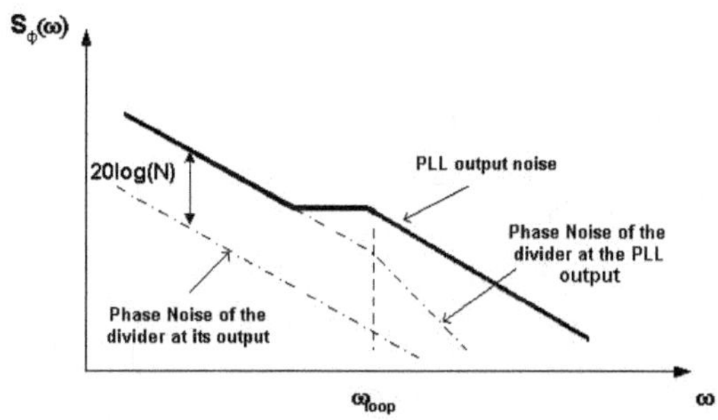

Fig. 10. Impact of the prescaler phase noise on the phase noise of a first-order loop with a low noise input.

The phase noise of a first-order PLL with a low noise input is depicted in Fig.10 for the sake of illustrating the effect of the divider phase noise on the output noise of the overall PLL system, [30]. As it is shown in the figure, the phase noise at the output of the divider is multiplied by the square of the divide ratio N when translated to the PLL output phase noise. Note that, even though the divider phase noise at offset frequencies larger than the loop bandwidth is suppressed by the loop, the divider phase noise at offsets lower than the loop bandwidth, denoted as ω_{loop}, contributes to the overall phase noise of the PLL. In the state-of-the-art frequency synthesizers, the loop bandwidth values are usually at a few tens of kHz and most of the wireless standards have demanding phase noise specifications at offsets lower than a few tens of kHz (e.g., DCS 1800 specification of -80dBc/Hz at 1kHz). This clearly illustrates the need for designing the prescalers for low noise.

3. Fully-Integrated RF Bandpass Filters

The existence of large interferers, spurious tones, unwanted image and carrier frequencies as well as their harmonics in the wireless communication environment mandates the use of bandpass filters with high selectivity in the front-ends of the transceivers, [7]. In the state-of-the-art solutions adopted in the industry, bulky and expensive high quality off-chip passive filters, SAW filters are used due to the demanding dynamic range specifications of the standards at the RF spectrum. There is a strong motivation behind implementing the filters in fully-integrated form using relatively cheap mainstream CMOS technologies mainly from the low cost and small form factor points of view. An additional factor that may favor integrated filters is that the off-chip filters require impedance matching circuits to interface with the integrated building blocks, complicating the system design with respect to an integrated solution. The main difficulties involved in integrating the filters at GHz range is the stringent standard specifications for the selectivity and the dynamic range. Although some progress has been reported in the literature using Q-enhanced LC filters, the feasibility of structures designed with mainstream standard CMOS technologies for low power high performance bandpass filters at frequencies higher than 2GHz is still being investigated. This issue is addressed in the proposed research using a prototype designed and fabricated with a $0.35\mu m$ standard CMOS technology.

C. A Brief Literature Review

Recently, successful CMOS implementations of fully-integrated frequency synthesizers for wireless communication systems have been reported, [29], [31], [34], and [35]. Besides the architectural optimization used in the selection of the synthesizer topologies, the presented works show considerable effort put to improve the performance of the individual building blocks, especially the VCOs and the prescalers, to meet the target standard specifications.

Negative conductance LC VCOs are by far the most widely used VCO structures in GHz range designs. The simplicity of the structure that allows the generation of a differential negative conductance with only two transistors in the signal path makes the circuit very attractive for applications where low noise with low power consumption at GHz range are key performance metrics, as it is the case for the wireless communication systems design, [2]. The circuits proposed in [36], [37], [38], are all based on the compensation of the LC tank losses by the negative conductance generated through cross-coupled Bipolar devices. The structure presented in [37] uses external LC resonators with high quality factors (Q) to achieve low phase noise levels with lower power consumption. The emphasis is put on the design challenges involved in the automatic amplitude control embedded into the circuit to keep the oscillation amplitude constant. On the other hand, the works reported in [36] and [38] use fully-integrated LC tank elements, with the emphasis put on the design and optimization of the passive tank elements to improve the performance.

CMOS negative conductance LC VCO structures are proposed in [34], [39] and [40]. The circuits reported in [34] and [39] use complementary cross-coupled NMOS and PMOS pairs to compensate for the loss of the tank with less current consumption compared to the case of using only NMOS cross-coupled pair saving some power. The design of the integrated spiral inductor in the LC tank is an important aspect of the work presented in [34]. The inductor was optimized for high quality factor at the frequency of operation to reduce the noise contributed by the resonator. The varactor is the main issue in the VCO reported in [39]. A 9.8GHz CMOS VCO tuned through the bulk terminal of the cross-coupled PMOS pair in $0.35\mu m$ CMOS is presented. The capacitance in the LC tank is the total parasitic capacitance at the output nodes and it is varied by changing the DC voltage applied to the bulk-terminal of the PMOS pair. As a result, the tuning range is only 2.8% but the operating frequency is very high. Minimizing the upconversion of the device flicker noise around the carrier using some circuit balancing techniques is the subject of [40]. Two fully-integrated LC VCOs using NMOS cross-coupled devices and different tail current sources are compared with respect to their noise performances at close-in offsets from the carrier. The 2GHz VCOs were designed with a $0.65\mu m$ BiCMOS technology using only MOS devices. The inductors used in the designs were optimized with a dedicated inductor-simulator/optimizer for a high quality factor value of 11 at 2GHz. A linear time-variant phase noise theory is presented in [30], [41]. The theory identifies the importance of symmetry in the oscillating waveform in suppressing the upconversion of $1/f$ noise around the carrier and it proves useful to calculate the contributions of each component in a VCO to the phase noise, [37].

There is a continuing research effort to improve the performance of the prescalers designed with the mainstream CMOS and BiCMOS technologies. The most important design trade-off in the prescalers is the one between the frequency of operation and the power consumption. Researchers in the field struggle to design circuits that will operate at higher frequencies with lower power consumption so do the PLL based frequency synthesizers the circuits are designed for. A 1.75GHz divide-by-128/129 prescaler in 0.7μm CMOS is reported in [32]. The dual-modulus prescaler which divides the input frequency (the frequency of the signal generated by the VCO) either by 128 or 129 depending on its modulus control signal is configured in such a way that the high speed section of the prescaler is limited to only one divide-by-two flip flop. In this way, the prescaler can operate at the same speed as an asynchronous divider. The circuit uses a supply voltage of 3V and draws 8mA of current. The measured phase noise at the output of the divider with an input at 1.28GHz in divide-by-128 mode is -131dBc/Hz at 1kHz which corresponds to a phase noise contribution of $-131 + 10\log(128^2) = -88.85dBc/Hz$. The bias network of the fully-differential circuit is not integrated in the prototype and the blocks are supplied externally.

A fixed-mode divide-by-eight frequency divider that can be used as a prescaler is presented in [42]. The circuit was integrated in a 0.8μm Silicon Bipolar Technology. The measured results show operation up to 15GHz while 22mA is drawn from a 3.6V supply. The circuit has single-ended input and output but all the signal processing is done differentially in between. The flip-flops used in the design are based on emitter-coupled-logic (ECL) and they are of master-slave type. The circuit operates from around 500MHz up to 15GHz and the input sensitivity over the frequency range from 2GHz to 14GHz is around 20dB at room temperature. The measurements were taken on the packaged parts and the divider provides an output power of -7dBm to a 50Ω load with an input at 15GHz. The bias network is integrated and the chip measures 536 X 476 μm^2. The measured phase noise at 1kHz offset with an input at 12GHz is -137dBc/Hz. A truly modular and power-scalable architecture for low-power programmable frequency dividers is reported in [33]. The architecture was used to realize programmable divider circuits which could be used as prescalers around 900MHz and 1.7GHz. The technology used is a 0.35μm CMOS technology. The key circuits in the dividers used as modular building blocks are divide-by 2/3 cells. Source-coupled-logic (SCL) based structures were used in the implementation. The circuits were biased externally and current scaling was done in the design in accordance with the frequencies the blocks should operate at to minimize the current consumption. The supply voltage used is 2.2V and the divider at 1.8GHz draws around 2.3mA while the divider at 900MHz draws around 900μA. The circuits are fully-differential including the input amplifiers that should be driven differentially.

The strong need for filtering out the unwanted frequency components in transceiver front-ends to improve the quality of the communication, and the disadvantages involved in using discrete off-chip passive filters for this task mainly from the cost and form factor points of view urged the designers to seek for solutions to integrate continuous-time filters on silicon in the GHz range. A second-order active bandpass filter at 1.8GHz using integrated spiral inductors in a Si Bipolar technology was reported in [43]. The filter uses the Q-boosting technique with partial- positive feedback to increase and tune the quality factor from 3 to 350. The measured frequency tuning is from 1.6GHz to 2GHz and it is performed through reactance multiplication. The circuit operates within a supply voltage range between 2.8V and 4.5V drawing around 8.7mA. The 1dB compression point dynamic range at room temperature is reported as 40dB for a Q of 35 at 1.8GHz. In the 1.9GHz receiver designed with a 0.5μm Bipolar Technology presented in [44], an alternate method of using a notch filter is proposed to filter out the unwanted image frequency component. The frequency tuning required to center the notch is performed using on-chip varactors and a tuning between 2.34GHz and 2.55GHz is measured. The receiver front-end which includes an LNA, an image-reject notch filter and a Gilbert cell mixer draws a total of around 15.9mA from a 3V supply. The receiver is designed to operate with a 1.9GHz RF signal and a 2.2GHz local oscillator (LO) signal for a 300MHz IF in which case the image frequency is 2.5GHz. The Q of the notch is tunable and the measurements show a rejection of 65dB for the 2.5GHz image component. Design and performance issues of fully-integrated Q-enhanced LC bandpass filters are discussed in [45]. The paper also reports a fully-integrated 850MHz fourth-order bandpass filter in a 0.8μm CMOS technology. The frequency tuning is performed by switching grounded metal-metal capacitors in and out of the resonator circuits, while the Q-tuning is done using a similar technique presented in [43]. The circuit draws around 77mA from a supply voltage of 2.7V occupying a die area of $2mm^2$. The measured 1dB compression point dynamic range with a 1-MHz IF bandwidth is 75dB for a Q of 47 and the circuit achieves a rejection over 50dB at 100MHz offset from the center frequency. Two magnetically coupled identical Q-enhanced LC resonators form a balanced 4th-order bandpass filter in the work presented in [46]. Automatic tuning circuitry is integrated on-chip for both the Q and the frequency tuning. The system is designed for 1.9GHz PCS standard with a 0.25μm BiCMOS technology. The frequency tunability is provided by the varactors and once again tunable negative resistance cells are used for the Q enhancement. The center frequency of the filter is 1.88GHz with a bandwidth of 150MHz and the ultimate rejection is over 50dB. The 1dB compression point dynamic range with the noise integrated over a bandwidth of 1GHz is 63dB. The main filter draws 18mA from a supply voltage of 2.7V, while an additional current of 18mA is drawn by the tuning circuit. The die area is 3.4mm X 2.1mm.

D. Specifics of the Proposed Research

1. Negative-Conductance LC Voltage-Controlled Oscillators

The basic negative conductance VCO is depicted conceptually in Fig.11(a). The inductor and the variable capacitor form the resonator tank, and the conductance G_{loss} represents the loss due to the finite quality factor of the resonator.

The negative conductance is denoted as $-G$ in the illustration. A sample circuit implementation in CMOS is shown in Fig.11(b), where the differential oscillator is built with symmetry around the resonator tank. The negative conductance $-G$ generated by the cross coupled NMOS transistors M_1 and M_2 is designed to compensate for the loss associated with the LC tank. It can be shown that the negative-conductance seen differentially at the drain of the NMOS transistors is $-G \approx -g_m/2$, where g_m denotes the transconductance of each transistor. In order for the oscillations to be sustained in the circuit, the tail current of the cross coupled transistor pair should be increased to a level high enough for the negative conductance to be larger in absolute value than the equivalent loss of the tank.

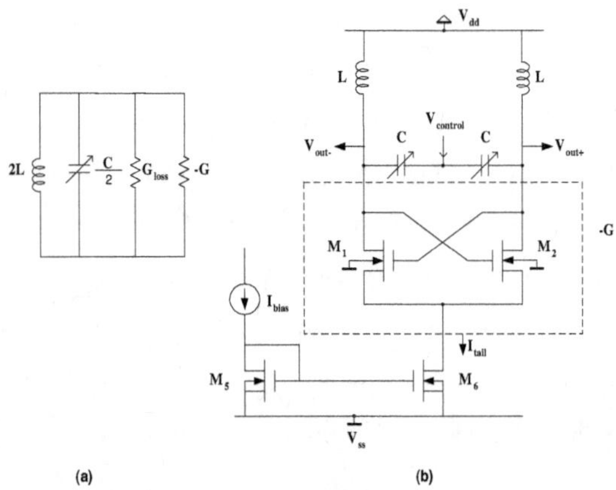

Fig. 11. Negative conductance voltage-controlled oscillator: (a) concept, (b) circuit implementation in CMOS.

The importance of the phase noise performance of the VCO was underlined in the first section. It is possible to calculate the phase noise of a VCO based on the impulse sensitivity function, $\Gamma(\omega_o\tau)$, [30], which is a dimensionless, frequency and amplitude independent function periodic in 2π that characterizes how sensitive the output phase of the oscillator is to an impulse injected into it. The phase noise of the symmetric linearized CMOS LC VCO presented in Chapter II is calculated using the impulse sensitivity function theory explained in [30] and the results are reported in detail in the associated section.

2. Dual-Modulus Prescaler Design in BiCMOS

The block diagram of the dual-modulus prescaler depicted for single-ended processing for the sake of clarity is shown in Fig.12. The input buffer functions as a single-ended to differential converter driving the synchronous divide by 4/5 counter. The output of the synchronous counter drives the divide by 8 extender which is comprised of three consecutive asynchronous divide by 2 stages. The additional logic in the feedback loop between the outputs of the asynchronous dividers and the synchronous divide by 4/5 counter is designed to modify the divide ratio in accordance with the *Mode* signal. If the *Mode* signal is at logic 0, than only the first two D Flip-Flops in the synchronous counter are involved in the loop dividing the input frequency by four and consequently the prescaler divides by 32. Otherwise, all the three D Flip-Flops in the synchronous counter are involved in the loop introducing an additional delay to divide the input frequency by 33. Once the RF input signal is converted into a differential signal, all the signal processing is done differentially to improve the immunity to common-mode noise sources such as the noise on the substrate and on the power supply lines. ECL-based D flip-flops and differential NOR gates are adopted in the design of the building blocks and the ECL signal level is set at around $200mVpp$.

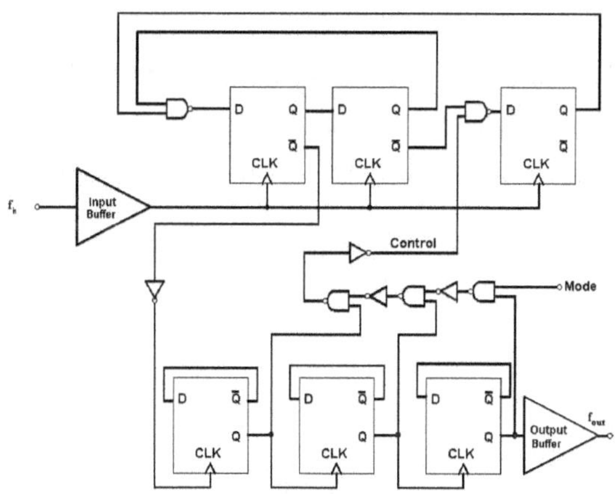

Fig. 12. Block diagram of the dual-modulus divide by 32/33 prescaler.

The use of ECL level signals in the digital building blocks proves advantageous over the use of CMOS levels in the integrated circuits which accommodate sensitive analog circuits on the same die from the point of view of reducing the disturbance to the sensitive analog signal lines. In the context of the PLL-based frequency synthesizer design, an example of such a sensitive analog signal line is the frequency control input line of the VCO. The switching noise coupled from the prescaler would disturb the VCO output significantly, creating spurs at the synthesizer output. The ECL-level signals would create less noisy disturbances on the substrate than the CMOS-level signals, providing more immunity to the VCO frequency control input.

The biasing of the prescaler is supplied through a bandgap bias generator from which all the bias currents of the building blocks are distributed. The phase noise contribution of the bias generator can be significant, requiring proper measures to be taken to design the circuit for low noise. The prescaler is designed to process a single-ended signal coming from an off-chip VCO. The input buffer acts like a single-ended to differential converter for this purpose. The output of the input buffer provides the ECL level differential clock signals to the synchronous divide-by-4/5 stage. The buffer is a simple differential pair followed by emitter followers to drive the resistive-capacitive loads in the synchronous divide-by-4/5 stage. Note that the buffer should be designed carefully to assure a reasonable input sensitivity with as low of a power consumption as possible.

3. Fully-Integrated LC Q-Enhancement RF Bandpass Filters in CMOS

The Q-enhancement LC filters can be interpreted as negative-conductance LC VCO based second-order bandpass filters which are obtained by moving the imaginary poles of the system to the left half-plane through introducing loss. Another interpretation is to use an on-chip LC tank for a second-order bandpass filter and to boost the Q of the filter by incorporating a negative-conductance (negative-resistance) generator.

The circuit implementation in CMOS is based on the use of the NMOS VCO core shown in Fig.11(b) as the load of the input transconductance stage. The frequency tuning is through the varactors whereas the Q is tuned by changing the tail current of the negative-conductance core. The circuit is designed and fabricated using a $0.35\mu m$ CMOS technology. The design equations of the Q-enhancement LC filter presented in Chapter IV demonstrate the strong effect of the starting quality factor of the tank on the performance of the filter. The design considerations, trade-offs and performance metrics (1dB compression point, noise figure, etc.) based on the simulations and theoretical calculations are presented in detail. The integrated circuit measurement results are provided together with some comparisons with the simulations. The advantages and disadvantages of the structure based on the measured data are evaluated.

CHAPTER II

NEGATIVE-CONDUCTANCE LC VOLTAGE-CONTROLLED OSCILLATORS

VCOs are among the key building blocks of wireless communication systems. The quality of the VCO is vital in setting the performance of the PLL-based frequency synthesizers used in generating the Local Oscillator (LO) signals that convert the transmitted and received information signals in frequency. The trend toward low cost and large scale integration in wireless communications systems design steadily increases the demand for fully integrated VCOs at RF in CMOS and BiCMOS technologies. Negative conductance based VCO structures using integrated spiral inductors and varactors as resonators show promising results as more and more efforts are put on improving the quality of the integrated resonator passive elements and the negative conductance generators by the process developers and the designers, [36], [47], [39], [48], [49], [40], [50], [38]. After briefly emphasizing the impact of the VCO quality on the performance metrics of PLL-based frequency synthesizers in Section A, design examples of negative conductance LC VCOs in Si BiCMOS and CMOS technologies are provided in Section B, where the advantages and shortcomings of the structures are presented together with the associated design trade-offs. Detailed design considerations in a symmetric linearized CMOS LC VCO with Bulk-Tuned PMOS capacitors is documented in Section C. Experimental results of all the CMOS VCOs are given in Chapter V.

A. VCO in a PLL

PLL is a widely used building block in modern communication systems, mostly in synthesizing well-defined, stable local oscillator signals for up and down-converting the transmitted and received signals, respectively, [29]. It is well accepted that the most challenging parts of especially high-frequency PLLs used in wireless communication systems are the VCOs due to the very stringent low noise specifications with low power consumption at GHz frequencies. The increasing demand for complete integration in relatively low cost IC technologies, i.e. CMOS, BiCMOS, without any performance compromise, as the frequency of operation increases steadily complicates the design even further.

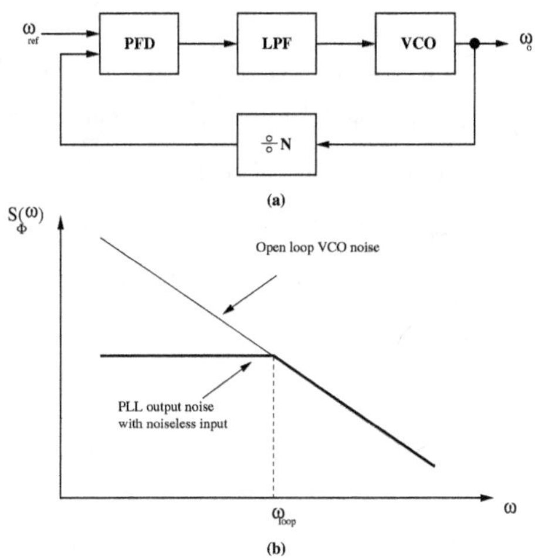

Fig. 13. Phase-locked loop: (a) basic block diagram, (b) phase noise of a first-order loop with noiseless input.

Block diagram of a basic PLL is shown in Fig.13(a). A reference signal usually from a low noise crystal oscillator is compared with the divided VCO signal by the Phase Detector (PD) and the resulting signal after low-pass filtered is connected to the VCO as the control voltage, closing the loop. When the loop is in locked condition, the output frequency of the PLL, i.e. the output frequency of the VCO, becomes equal to the reference frequency multiplied with the divider ratio, that is

$\omega_o = N \cdot \omega_{ref}$. The demanding phase noise specifications of wireless communication standards sets the major design challenge as achieving low noise at the PLL output. The phase noise of a first-order PLL with a noiseless input is depicted in Fig.13(b) for the sake of illustrating the effect of the VCO phase noise on the output noise of the overall PLL system, [30]. Note that, even though the VCO phase noise at offset frequencies lower than the loop bandwidth is suppressed by the loop, the phase noise of the PLL is determined by that of the VCO at offset frequencies larger than the loop bandwidth, denoted as ω_{loop}. On the other hand, the need for synthesizing frequencies with small steps limits the loop bandwidth to small values, [29], thus the VCO phase noise remains unsuppressed at the output of the PLL for some large offset frequencies where the specifications of wireless standards are demanding. This clearly shows the importance of low noise VCO design for a high-performance frequency synthesizer complying with the standards.

B. Negative Conductance Voltage-Controlled Oscillators

Negative conductance voltage controlled oscillators are by far the most widely used VCO structures in GHz range designs. The simplicity of the structure that allows the generation of a differential negative conductance with only two transistors in the signal path makes the circuit very attractive for applications where low noise with low power consumption at GHz range are key performance metrics.

The basic negative conductance VCO is depicted conceptually in Fig.14(a). The inductor and the variable capacitor form the resonator tank, and the conductance G_{loss} represents the loss due to the finite quality factor of the resonator. The negative conductance is denoted as $-G$ in the illustration. A sample circuit implementation in CMOS is shown in Fig.14(b), where the differential oscillator is built with sym-

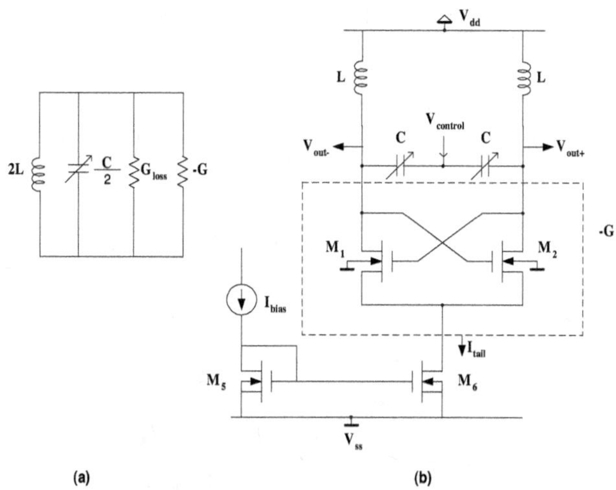

Fig. 14. Negative conductance voltage-controlled oscillator: (a) concept, (b) circuit implementation in CMOS.

metry around the resonator tank. The negative conductance $-G$ generated by the cross coupled NMOS transistors M_1 and M_2 is designed to compensate for the loss associated with the LC tank. It can be shown that the negative-conductance seen differentially at the drain of the NMOS transistors is $-G \approx -g_m/2$, where g_m denotes the transconductance of each transistor. In order for the oscillations to be sustained in the circuit, the tail current of the cross coupled transistor pair should be increased to a level high enough for the negative conductance to be larger in absolute value than the equivalent loss of the tank.

1. Bipolar VCO Design

In this section, some design trade-offs involved in the design of Bipolar negative conductance VCOs are briefly provided. The schematic of a Bipolar VCO is shown in

Fig.15. The cross coupled transistors Q_1 and Q_2 generate the negative conductance required to compensate for the loss associated with the finite quality factor of the LC tank which sets the frequency of oscillation.

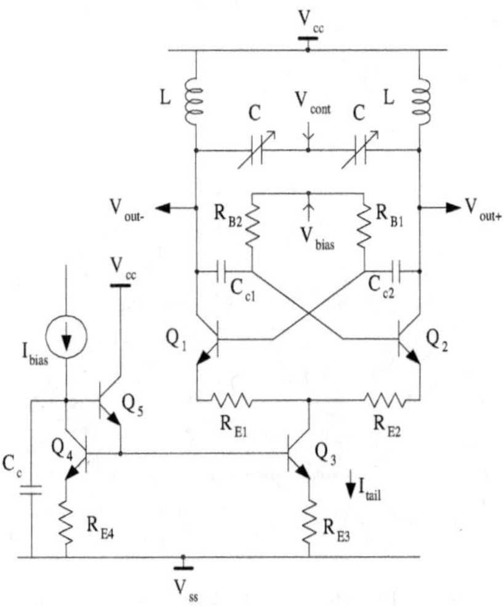

Fig. 15. Bipolar voltage-controlled oscillator.

The tank varactor is a reverse-biased PN junction available in the 0.6 μm BiC-MOS technology used in the design. It can also be the base-emitter or base-collector junctions of a bipolar transistor used in reverse-bias, [38]. The emitter-degeneration resistors R_{E1} and R_{E2} are used to reduce the upconversion of noise around the carrier at low offsets by linearizing the negative conductance generator, [36]. Care should be taken in optimizing the value of the resistors through phase noise simulations as their thermal noise contribution at large offsets may degrade the phase noise. Another important concern is that the addition of the degeneration resistors reduces the loop gain. Thus the start-up of the oscillations should be guaranteed over the process variations. The capacitive coupling through C_{c1} and C_{c2} in the feedback of the cross coupled transistors Q_1 and Q_2 is used to avoid the clipping at the output swing due to the saturation of the transistors. The simulations show that the voltage swing at the output of the VCO is almost doubled with capacitive coupling as compared to direct coupling. As a consequence, the phase noise is improved considerably since it is inversely proportional to the voltage swing,[30]; more than 8 dB improvement at 600kHz offset is observed in the SpectreRF simulations of the VCO designed to operate at around 2.5 GHz. Note that, capacitive coupling also reduces the loop gain through capacitive division. Careful design of the capacitive coupling is therefore vital to ensure the oscillations. It is also important to minimize the phase noise contribution from the bias resistors R_{B1} and R_{B2}, [38]. The cross coupling in the negative conductance generating transistors can also be done through emitter followers in lieue of the capacitors, [36]. Although the resulting structure also proves useful in preventing the amplitude clipping that occurs due to the forward biased base-collector junction, the additional power consumed by the emitter followers and their phase noise contribution may be a concern. The biasing of the VCO core shown in Fig.15 is supplied by a bandgap bias generator that feeds the tail current mirror. The phase noise contribution of the bias generator through upconversion of low frequency noise around the carrier can be significant, requiring proper measures to be taken to design the bias circuit for low noise.

2. CMOS VCO Design

The emphasis in this section is put on analytically calculating the phase noise of a CMOS VCO designed with HP 0.5μm CMOS Technology, based on a time-variant phase noise theory, [41]. The theory proves useful to calculate the contributions of each component in a VCO to the phase noise, [37], thus allowing for optimization in the design process. Schematic of the CMOS VCO with complementary cross coupled transistors around an LC tank is shown in Fig.16.

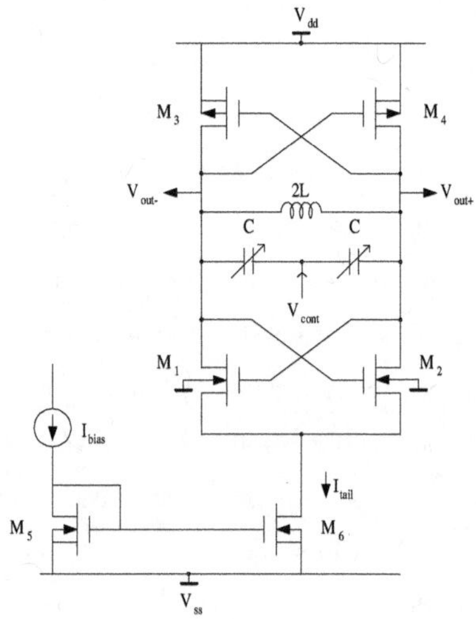

Fig. 16. CMOS voltage-controlled oscillator.

The negative conductance generating cross coupled NMOS and PMOS transistors, M_1-M_4 are designed to compensate for the loss associated with the LC tank. The advantage of using both PMOS and NMOS cross coupled transistors is mainly two-fold: first, with the addition of the PMOS cross coupled pair, it is possible to compensate for the loss of the tank with less current consumption saving some power, [34], and second by observing the symmetry properties of the oscillating waveform sizing the PMOS and NMOS transistors ($g_{mp} = g_{mn}$), it is possible to reduce the up-conversion of $1/f$ noise of the transistors around the carrier, thus lowering the phase noise, [30]. On the other hand, it is obvious that the additional headroom required for the biasing of the PMOS transistors as compared to the NMOS-only structure shown in Fig.14(b), makes the complementary structure more difficult to design with low supply voltages.

The tank inductor is a spiral structure. The component values in the simplified Pi model of the spiral inductor were taken from [51]. AC simulation of the inductor model predicts a quality factor of around 2.5 at 2GHz. It is an octagonal structure built with three metal layers of the technology, connected in series. The cross coupled negative conductance generator is to be connected to the third metal layer to minimize the capacitive coupling to the substrate. The varactor is a PMOS capacitor operating in depletion. The phase noise of the CMOS VCO is calculated based on the impulse sensitivity function, $\Gamma(\omega_o\tau)$, [30], which is a dimensionless, frequency and amplitude independent function periodic in 2π that characterizes how sensitive the output phase of the oscillator is to an impulse injected into it. Note that, as the value of the impulse sensitivity function depends on at what phase, $\omega_o\tau$, the impulse is injected into the oscillator, time-variance is involved into the computations. The phase noise spectrum in the $1/f^2$ region is given by, [41]

$$\mathcal{L}\{\Delta\omega\} = \sum_n \frac{\frac{\overline{i_n^2}}{\Delta f}\Gamma^2_{rms,n}}{2q^2_{max}\Delta\omega^2} \tag{2.1}$$

where $\Delta\omega$ is the offset frequency from the carrier, $\Gamma_{rms,n}$ is the rms value of the impulse sensitivity function (ISF) of the nth noise-generating component of the oscillator, $\overline{i_n^2}/\Delta f$ is the mean-square noise current density of the associated component and q^2_{max} is the maximum charge swing across the tank capacitance. As the impulse sensitivity function is periodic, it can be expressed as a Fourier Series

$$\Gamma(\omega_o\tau) = \frac{c_0}{2} + \sum_{n=1}^{\infty} c_n \cos(n\omega_o\tau) \tag{2.2}$$

where the coefficients c_n are real. c_0, the dc coefficient of ISF governs the upconversion of $1/f$ device noise around the carrier, whereas the other coefficients are responsible from the downconversion of white noise near the harmonics of the carrier, [30], [40]. The spectrum of phase noise in the $1/f^3$ region, where the upconversion of device $1/f$ noise around the carrier is observed is described by, [41]

$$\mathcal{L}\{\Delta\omega\} = \sum_n \frac{\frac{\overline{i_n^2}}{\Delta f} c_0^2}{8q_{max}^2 \Delta\omega^2} \cdot \frac{\omega_{1/f}}{\Delta\omega} \tag{2.3}$$

where $\omega_{1/f}$ denotes the $1/f$ noise corner of the associated device in the oscillator. As the equations describing the spectrum of phase noise show, the ISF of all the noise-generating components should somehow be calculated to compute their contribution to the phase noise of the VCO. Once the individual contributions are computed, the phase noise of the VCO can be determined through summation as (2.1) and (2.3) suggest. Note that each $\overline{i_n^2}/\Delta f$ in the equations represents drain current noise, resistor noise, varactor noise and inductor noise. The varactor and inductor noise are due to the finite quality factor, Q, of the components and they can easily be expressed in terms of the loss resistances of the components which can be obtained from their respective simulated Q values at the frequency of oscillation.

To determine the ISF of a noise-generating component in the VCO, an impulsive current source that represents the associated noise source was connected in parallel with the component. The resulting phase shift was measured through transient simulations, as explained in [30]. The area of the current impulse injected corresponds to the amount of charge perturbation applied to the circuit and in order for linearity, which is one of the bases of the theory, to hold, the amount of charge injected should be much smaller than the steady-state charge swing across the tank capacitance. A scaling on the measured phase shift was done after the transient simulations, accordingly. As mentioned before, the value of ISF, $\Gamma(\omega_o\tau)$, depends on at what phase,

$\omega_o\tau$, the current impulse is injected into the circuit and it is this characteristic of the ISF that accounts for the time-variance involved in the theory. In order to determine the Fourier coefficients of the ISF of a noise-generating component accurately, the number of current impulse injection simulations within an oscillation period, which corresponds to 2π radians, should be kept as high as possible. The current impulse is injected once the oscillating waveform reaches its steady-state. The simulations are repeated N times by changing the phase at what the current impulse is injected within one oscillation period with $2\pi/N$ increments each time with respect to the starting phase. Obviously, the accuracy with which (2.1) and (2.3) estimate the phase noise depends on the accuracy with which the Fourier coefficients are determined. As an approximate figure for the phase noise of the VCO was targeted for the analytical calculations, the mean-square value of the ISFs of the tank inductor and the tank varactor were taken as 1/2, as is the case for an ideal sinusoidal waveform of an LC oscillator, [30]. Furthermore, the number of transient simulations for the ISF determination of each component was kept at a relatively small number of 50.

A comparison between the analytically calculated and simulated phase noise of the CMOS VCO is plotted in Fig.17. The calculated phase noise agrees well with the simulation result. The better matching at the $1/f^3$ spectrum of the phase noise indicates that the accuracy in the determination of the dc values of the ISFs of the components is somehow superior to the accuracy in the determination of the mean-square values of the ISFs of the components, through charge injection simulations.

The calculated percentage contributions at 3MHz offset show that, the low Q inductor is the main contributor to the phase noise at large offset frequencies. At low offsets from the carrier though, i.e. at 1kHz, with the addition of the upconverted flicker noise of the transistors, the percentage contributions differ considerably. The largest contributor is the NMOS cross coupled pair, while the PMOS cross coupled

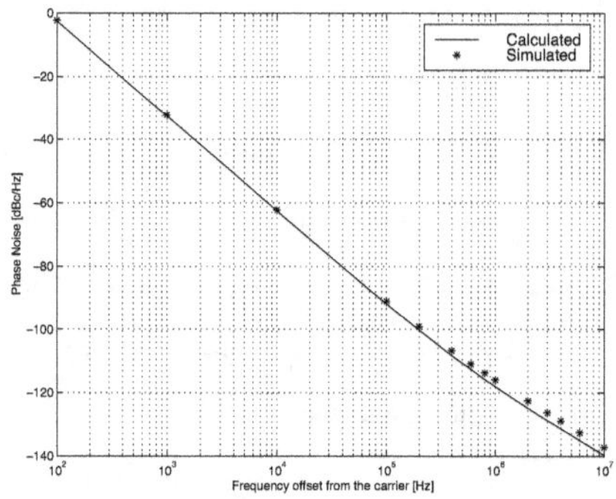

Fig. 17. SpectreRF simulated vs. analytically calculated phase noise of the CMOS VCO.

pair contributes much less with its superior flicker noise in the technology. By observing the symmetry properties of the oscillating waveform through the proper sizing of the cross- coupled transistors it is possible to reduce the dc values of the ISFs of the MOS transistors, which govern the upconversion of the flicker noise around the carrier, to improve the close-in phase noise, [30]. As for the phase noise at large offsets, an integrated LC tank with higher quality factor should be used.

A second VCO designed with only NMOS transistors in the negative-conductance generator uses the same inductor as its complementary cross coupled counterpart. The same PMOS varactors are used in the tank. The NMOS-only VCO achieves nearly the same phase noise level as the VCO with the complementary cross coupled pairs investigated above but with almost twice the power consumption. The lack of the PMOS pair prevents the optimization of the negative-conductance generator transistors for lower noise upconversion. The advantage of the NMOS-only VCO is that the circuit can operate with a lower supply of 2.5V as opposed to the 3V supply of the complementary one. As the Q of the inductor used in the tank is around 2.5

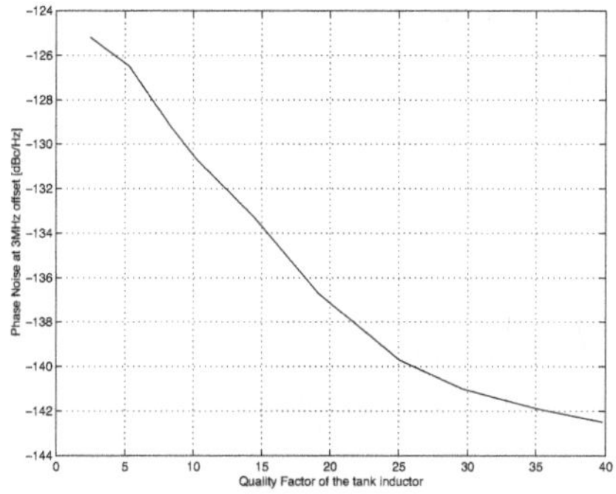

Fig. 18. Simulated phase noise of the NMOS-only VCO @ 3MHz offset vs. the quality factor of the tank inductor.

and the Q of the varactor is around 35 at the frequency of oscillation, the quality factor of the tank is determined by that of the inductor. In order to evaluate how the quality factor of the inductor affects the phase noise of the VCO at large offsets quantitatively, SpectreRF simulations were performed using the NMOS-only VCO test bench, by setting the Q of the tank inductor as a parameter. The simulation results are plotted in Fig.18, where the y-axis is the simulated phase noise at 3 MHz offset in dBc/Hz and the x-axis is the Q of the inductor. A phase noise value of around -131dBc/Hz is obtained with an inductor Q value of 10, which is a realistic Q value for optimized integrated inductors, [40]. The significant improvement in the phase noise performance as the inductor Q increases as observed in Fig.18 clearly illustrates the dominant contribution of the inductor noise in the phase noise of the VCO at large offsets.

C. Design Considerations in a Symmetric Linearized CMOS LC VCO with Bulk-Tuned PMOS Capacitors

RF CMOS is dominating the research activities in IC design for wireless communication transceivers. The ever increasing interest in the industry for using CMOS in RF, mostly due to integrability with the CMOS DSP circuitry, causes most of the research effort to be concentrated on exploiting the use of CMOS circuits for RF functions. In this work, the design considerations of a symmetric linearized CMOS LC VCO are investigated. Discussions on the optimization and design issues of generic LC VCOs are available elsewhere, [52]-[56]. PMOS capacitors tuned from the bulk (back gate) are used as tank varactors in the VCO. The frequency tuning limitation due to the well resistance is pointed out and experimentally verified. The impact on the phase noise of linearizing the negative resistance generator using source-degeneration resistors is demonstrated. The technique of using emitter-degeneration resistors to reduce the upconversion of device noise around the carrier through linearization is used in Bipolar VCOs, [36], [57], but no investigation on using source-degeneration resistors in the CMOS cross-coupled LC VCOs was reported to-date, in the literature known to the author.

In the following section, the symmetric linearized CMOS VCO is introduced. The design issues involved in the use of bulk-tuned PMOS capacitors and the source-degeneration resistors are investigated from the perspective of tuning range and phase noise. The Linear Time-Variant Impulse Sensitivity Function theory was used to calculate the phase noise contributions of the components in the VCO to gain insight about the impact of the different components to the phase noise at low and large offsets from the carrier. The measurement results of the prototype VCO are provided in Chapter V.

1. CMOS Voltage-Controlled Oscillator Design

The basic VCO concept is shown in Fig.19(a). The circuit is a differential oscillator with full symmetry around the integrated tank. The negative resistances ($-R_N$ and $-R_P$), shown in Fig.19(b), generated by the cross-coupled NMOS and PMOS transistors respectively are designed to compensate for the loss associated with the LC tank, which is denoted as R_{loss} in Fig.19(a). The advantage of using complementary cross-coupled transistors is mainly two-fold: first, with the addition of the PMOS pair, it is possible to compensate for the loss of the tank with less current consumption saving some power, [34], and second by observing the symmetry properties of the oscillating waveform matching the PMOS and NMOS transistors ($g_{mp} = g_{mn}$), it is possible to reduce the upconversion of $1/f$ noise of the devices around the carrier, thus lowering the phase noise, [55]. It is pointed out that the structure with complementary cross-coupled transistors provides almost twice the signal swing with the same tail current and the same LC tank with respect to the structure with NMOS-only cross-coupled transistors, [56], [58]. The importance of larger signal swing, thus larger signal power in the oscillator can be appreciated with the phase noise expression published by Leeson [59],

$$\mathcal{L}\{\Delta\omega\} = F\frac{kT{\omega_o}^2}{2P_{sig}Q^2\Delta\omega^2} \tag{2.4}$$

where P_{sig} denotes the signal power of the oscillator, Q is the quality factor of the tank, F is the noise factor of the active devices, ω_o is the carrier frequency, $\Delta\omega$ is the offset frequency from the carrier, k and T are the Boltzmann constant and the absolute temperature, respectively. Equation (2.4) approximates the phase noise in the $1/f^2$ region. Note that the lower the quality factor of the tank (Q), the larger the loss in the tank (R_{loss} in Fig.19(a)) and consequently, the greater the noise contributed by the tank, as it will be elaborated in Section II-C. Furthermore, the negative resistance generator designed to compensate for the loss of the tank would also contribute to the phase noise of the VCO, as represented by F in (2.4).

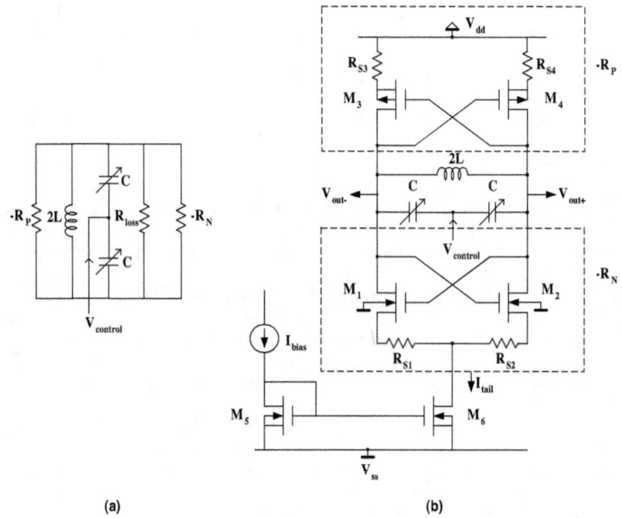

Fig. 19. CMOS voltage-controlled oscillator: (a) concept, (b) circuit implementation.

The source-degeneration resistors in the cross-coupled pairs of the CMOS VCO are added to reduce the flicker noise upconversion through linearizing the negative resistance by attenuation. As the addition of the resistors would modify the equivalent transconductances of the cross-coupled structures, the condition for the oscillating waveform symmetry (for R_{S1}=R_{S2} and R_{S3}=R_{S4}) mentioned above yields

$$\frac{g_{m1}}{1+g_{m1}R_{S1}} = \frac{g_{m3}}{1+g_{m3}R_{S3}}. \tag{2.5}$$

Equation (2.5) shows that the addition of the degeneration resistors provides additional design degrees of freedom in the architecture. Besides, the resistors reduce the differential negative conductance generated by the cross-coupled transistors by a factor of $(1 + g_m R_S)$ as expressed below

$$G \simeq -\frac{g_m}{2\left(1 + g_m R_S\right)}. \tag{2.6}$$

Therefore, the source-degeneration resistors would cause a power consumption penalty with the same LC tank compared to using a cross-coupled pair without degeneration resistors. The impact of the source-degeneration resistors on the phase noise is investigated in detail in Section II-B.

a. Spiral Inductors

The integrated spiral inductors used in the CMOS VCO are built with three metal layers of the technology, connected in series. The inductor model [51] used in the simulations is shown in Fig.20. It is a simplified asymmetrical model. Note that two inductors are connected back-to-back. The cross-coupled negative resistance generator is connected to the third metal layer of each inductor to minimize the capacitive coupling to the substrate. The port where the negative resistance generator is connected and the oscillating signal swings, is modeled with a capacitor in series with a resistor. The capacitor represents the capacitance between the third metal layer and the substrate and the resistor represents the substrate resistance. As the other port of the inductor is a common-mode node, i.e. AC ground, in the circuit, no such modeling is done at that port, for the sake of simplicity. AC simulation of the inductor model predicts a quality factor of around 2.5 at 2GHz for the inductors of 3.7nH each, as shown in Fig.21.

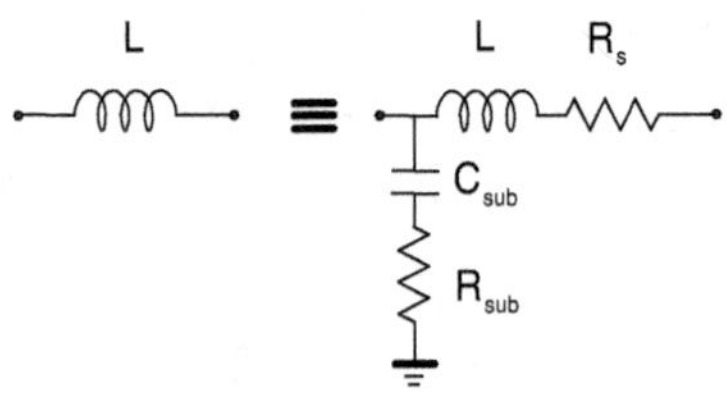

Fig. 20. Asymmetrical model of the integrated spiral inductor used in the differential VCO simulations.

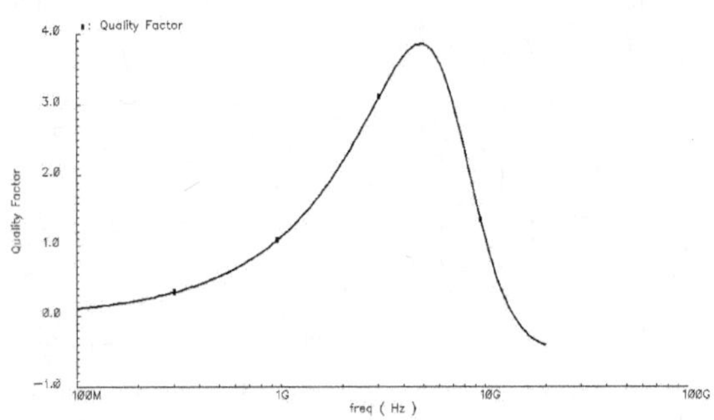

Fig. 21. Simulated spiral inductor quality factor vs. frequency.

b. Bulk-tuned PMOS Capacitors

PMOS capacitors tuned from the bulk are used as tank varactors in the CMOS VCO. The structure is a standard PMOS transistor available in digital CMOS technologies and the configuration used is depicted in Fig.22, where R_W shown in the figure explicitly for illustrative purposes denotes the well resistance. The oscillating voltage signal, V_{osc}, swings at the gate of the capacitor, whereas the control voltage is applied to the bulk. The idea of tuning the VCO through the bulk (back-gate) was reported in [39], where the back-gates of the cross-coupled PMOS transistors were used to tune

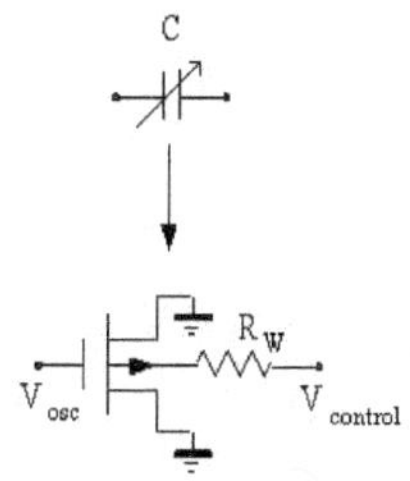

Fig. 22. PMOS bulk-tuned capacitor as tank varactor.

the frequency. In the present case, explicit bulk-tuned varactors are used to increase the tuning range and a limitation due to the well resistance is observed and reported. The effect of the well resistance on the tuning range is not mentioned in [39]. Note that the well resistance appears in series with the tuned capacitance in the bulk-tuned structure depicted in Fig.22, thus reducing the tuning range. The reduction in the tuning range is verified with the simulations. Incorporating an approximate calculated figure for the well resistance into the structure shown in Fig.22 shows a reduction in the capacitor tuning range from a ratio of 2.5:1 to 1.9:1 as depicted in the simulation result given in Fig.23. The corresponding reduction in the tuning range of the CMOS VCO in the simulations including some interconnection parasitics is from around 9.3% to 6.1%.

The bulk-tuned varactor operates in accumulation and depletion regions within the tuning range. As the drain and source terminals of the structure are connected to ground and the gate is biased at a positive voltage, no inversion occurs at the channels of the PMOS varactors for any bulk voltage within the tuning range. The quality factor of the bulk-tuned PMOS capacitor is changing from 8 to 40 within the tuning range from the accumulation to the depletion, in the AC simulations where both the well and the gate resistances are explicitly modeled. The change in the quality factor

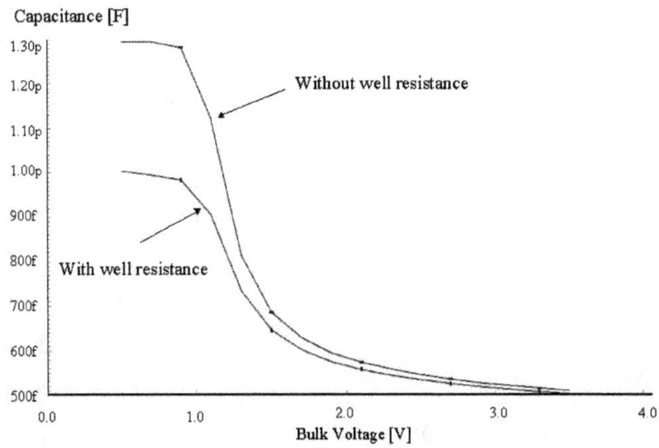

Fig. 23. Simulated effect of the well resistance of the PMOS bulk-tuned capacitor on the tuning.

of the capacitor with the addition of the well resistance is simulated over the tuning range. The result shown in Fig.24 clearly depicts the considerable degradation in the quality factor with the addition of the well resistance into the model. Note that the x-axis is the bulk voltage while the gate voltage is kept constant at the dc operating point of the output of the VCO.

This section is concentrated on deriving an analytical expression of the capacitance-voltage transfer function of the bulk-tuned PMOS capacitor used in the CMOS VCO. Numerous technical papers exist in the literature which deal with the design issues of integrated varactors to be used in RF integrated circuits for wireless communications, [60]-[66]. A comprehensive discussion is given in [64] regarding an analytical model for the MOS varactor transfer characteristic. The equations presented in the paper are based on the physical MOS capacitor model [67], [68]. A similar, yet rather limited discussion is also included in [61], the model of which is based on the treat-

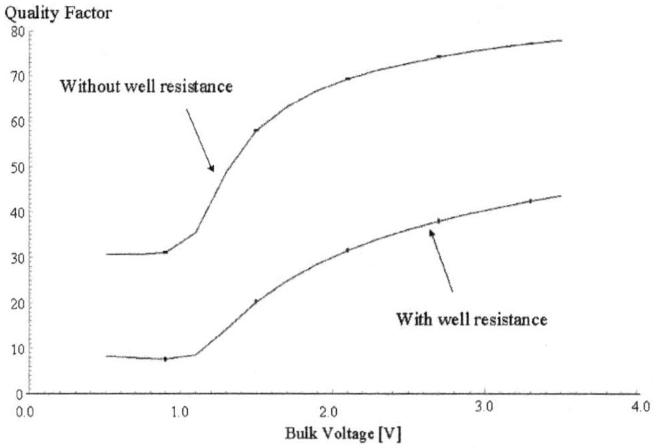

Fig. 24. Simulated effect of the well resistance of the PMOS bulk-tuned capacitor on the quality factor.

ment presented in [69]. In this work, related equations in [68] were used to derive an analytical expression of the capacitance-voltage transfer function of the bulk-tuned PMOS capacitor used in the CMOS VCO. It should be noted that, as the drain and source terminals of the structure are connected to ground and the gate is biased at a positive voltage as can be seen from the schematic shown in Fig.22, no inversion occurs at the channels of the PMOS varactors for any bulk voltage. This is due to the fact that for all the holes that exist in the bulk, the drain and source regions biased at a lower potential than the gate are more attractive than the surface for all possible gate-bulk voltages. The very same reason prevents the majority carrier holes in the p^+ doped drain and source regions to be attracted to the surface for inversion. The physical basis briefly explained above is taken into account in deriving the analytical expressions. A simplified small-signal equivalent circuit of the bulk-driven varactor structure based on the foregoing discussion can be depicted as shown in Fig.25.

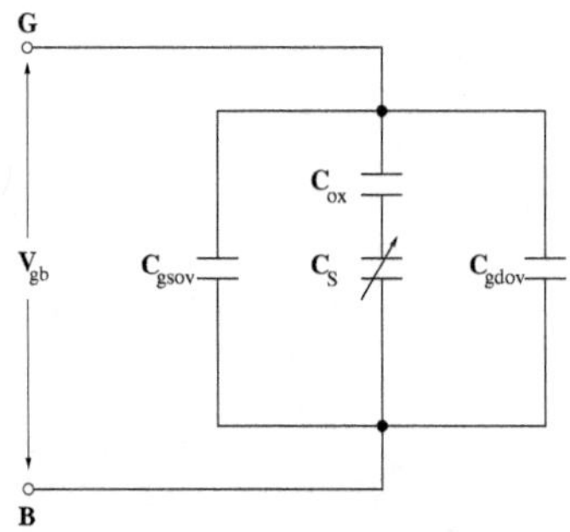

Fig. 25. Small-signal equivalent circuit of the capacitive component of the bulk-tuned PMOS varactor.

Based on the small-signal equivalent circuit of Fig.25, the total capacitance of the PMOS capacitor looking from the gate terminal, C_{gb}, as used in the VCO can be computed as follows:

$$C_{gb} = \frac{C_{ox} \cdot C_s}{C_{ox} + C_s} + C_{gsov} + C_{gdov} \tag{2.7}$$

where C_{gsov} and C_{gdov} denote the gate-source and gate-drain overlap capacitances, respectively. C_{ox} and C_s denote the gate oxide capacitance and the nonlinear semiconductor capacitance. Note that the semiconductor capacitance C_s is providing the variable capacitance of the structure. The gate-source and gate-drain overlap capacitances, C_{gsov} and C_{gdov}, are computed using the BSIM3 model parameters $CGSO$ and $CGDO$ with the equations given below:

$$\begin{aligned} C_{gsov} &= CGSO \cdot W \\ C_{gdov} &= CGDO \cdot W \end{aligned} \tag{2.8}$$

where W denotes the width of the PMOS capacitor. C_{ox} is given by

$$C_{ox} = \frac{\epsilon_{ox}}{T_{ox}} \cdot W \cdot L_{eff} \tag{2.9}$$

where ϵ_{ox} is the dielectric constant of Si_2O, T_{ox} is the gate oxide thickness and L_{eff} is the effective channel length of the PMOS capacitor. The nonlinear semiconductor capacitance C_s can be expressed as a function of the surface potential, Ψ_s, [68], [67]. In order to derive an expression for the semiconductor capacitance, total semiconductor charge per unit area should first be expressed as a function of the surface potential. Following the derivation procedure presented in [68], [67] and making the necessary modifications according to the characteristics of the structure used, the semiconductor charge per unit area can be expressed as

$$Q_s' = -\text{sign}\Psi_s \cdot \sqrt{2q\epsilon_s N_{sub}} \cdot \sqrt{\phi_t \cdot e^{\Psi_s/\phi_t} - \Psi_s - \phi_t} \tag{2.10}$$

where ϵ_s is the dielectric constant of silicon, q is the electron charge, N_{sub} is the doping concentration of the well and ϕ_t is the thermal voltage. Note that given a PMOS capacitor, $\Psi_s > 0$ corresponds to accumulation, whereas $\Psi_s < 0$ corresponds to depletion. By definition, semiconductor capacitance per unit area can be obtained by taking the derivative of (2.10) with respect to the surface potential according to the formula $C_s' = -dQ_s'/d\Psi_s$. Thus, one can express the semiconductor capacitance per unit area as a function of the surface potential as shown below:

$$C_s' = \text{sign}\Psi_s \cdot \sqrt{2q\epsilon_s N_{sub}} \cdot \left\{ \frac{e^{\Psi_s/\phi_t} - 1}{2\sqrt{\phi_t \cdot e^{\Psi_s/\phi_t} - \Psi_s - \phi_t}} \right\}. \tag{2.11}$$

The total semiconductor capacitance C_s is obtained by multiplying C_s' with the gate area of $W \times L_{eff}$. Having found all the components shown in the small-signal equivalent circuit of Fig.25, the total capacitance of the varactor looking from the gate terminal, C_{gb}, as used in the VCO can be computed as follows:

$$C_{gb} = \frac{C_{ox} \cdot C_s}{C_{ox} + C_s} + C_{gsov} + C_{gdov}. \tag{2.12}$$

Note that all the equations derived above express the capacitances as functions of the surface potential. In order to obtain the voltage-capacitance transfer characteristic of the PMOS varactor, the voltage accross the terminals of the varactor, which is the gate-bulk voltage V_{gb} should be expressed as a function of the surface potential. Once again following the derivation procedure presented in [68] and making the necessary modifications according to the characteristics of the structure used, the gate-bulk voltage can be expressed as a function of the surface potential as follows:

$$V_{gb} = V_{FB} + \psi_s + \text{sign}\Psi_s \cdot \gamma \cdot \sqrt{\phi_t \cdot e^{\Psi_s/\phi_t} - \Psi_s - \phi_t} \tag{2.13}$$

where V_{FB} is the flat-band voltage and γ is the body effect coefficient. The body effect coefficient can be calculated using the formula given below:

$$\gamma \equiv \frac{\sqrt{2q\epsilon_s N_{sub}}}{C'_{ox}} \tag{2.14}$$

where C'_{ox} denotes the gate oxide capacitance per unit area and is given by ϵ_{ox}/T_{ox}. It should be noted at this point that the analytical model presented so far is somewhat simplified mainly due to the fact that taking the doping concentration of the well, N_{sub}, as a constant, the effect of nonuniform doping along the channel is neglected, [64]. Despite the simplification, the model seems to match the simulation results within a maximum of 12% deviation. The flat-band voltage, V_{FB}, on the other hand, can be computed as follows:

$$V_{FB} = V_{TH0} - \phi_0 + \gamma \cdot \sqrt{-\phi_0} \tag{2.15}$$

where V_{TH0} is the zero-bias threshold voltage and ϕ_0 is the surface potential of the MOS structure in strong inversion taken as $\phi_0 = 2\phi_F$, [68]. Note that ϕ_F denotes the well Fermi potential and can be computed as

$$\phi_F = -\phi_t \ln \frac{N_{sub}}{n_i} \tag{2.16}$$

where n_i is the intrinsic carrier density of silicon. The model equations derived above can be used to calculate the gate-bulk capacitance of the PMOS bulk-tuned varactor for a given range of surface potentials. Then (2.13) can be used to compute the corresponding gate-bulk voltages to finally obtain the control voltage vs. capacitance transfer characteristic of the PMOS bulk-tuned varactor. A program was written in MATLAB for obtaining the transfer function using the analytical model derived above. A comparison between the simulated varactor transfer characteristic and the analytically computed varactor transfer characteristic based on the model equations derived is shown in Fig.26. Despite the simplifications used in the derivations, the model seems to match the simulation results within a maximum of 12% deviation.

A possible way to alleviate the limiting effect of the well resistance on the tuning range observed in the simulation results is to connect both the drain and the source terminals of the transistor used as the varactor to its well and to tune the capacitance from this node. In this configuration the well resistance is shunted by the much lower drain and source resistances. The capacitance of the structure is tuned by modifying the inversion level in the channel with the majority carriers in the source and drain regions through changing the control voltage. The inversion mode capacitor (varactor) obtained in this way is used in the frequency tuning of the RF filter to be presented in Chapter IV.

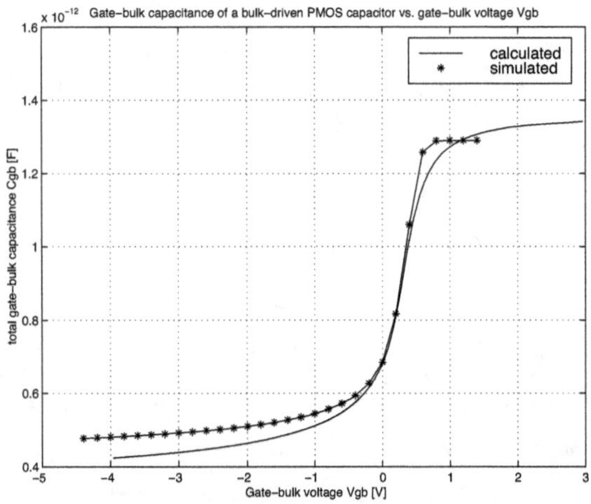

Fig. 26. Comparison between the simulated and the analytically calculated transfer characteristic of the bulk-tuned PMOS capacitor.

c. Phase Noise and the Source-Degeneration Resistors

The phase noise of the CMOS VCO with source degeneration resistors is analytically calculated based on the ISF based phase noise theory, [30]. As mentioned before, the theory proves useful to calculate the contributions of each component of the VCO to the phase noise, thus enabling the designer to optimize the main noise contributors for low noise design, [55], [37]. A similar approach where the phase noise contributions of each component in the oscillator were calculated was recently presented in [70]. The phase noise contribution of a noise-generating component in the VCO is calculated based on its impulse sensitivity function (ISF), $\Gamma(\omega_o\tau)$, [71], which is a dimensionless, frequency and amplitude independent function periodic in 2π that characterizes how sensitive the output phase of the oscillator is to the noise perturbations of the associated component. Note that, as the value of the impulse sensitivity function depends on at what phase, $\omega_o\tau$, the impulse is injected into the oscillator, time-variance is involved into the computations. For the sake of completeness, the equations used in calculating the component phase noise contributions are repeated in the following. The phase noise spectrum in the $1/f^2$ is given by, [41]

$$\mathcal{L}\{\Delta\omega\} = \sum_n \frac{\frac{\overline{i_n^2}}{\Delta f}\Gamma_{rms,n}^2}{2q_{max}^2\Delta\omega^2} \tag{2.17}$$

where $\Delta\omega$ is the offset frequency from the carrier, $\Gamma_{rms,n}$ is the rms value of the impulse sensitivity function (ISF) of the nth noise-generating component of the oscillator, $\overline{i_n^2}/\Delta f$ is the mean-square noise current density of the associated component and q_{max}^2 is the maximum charge swing across the tank capacitance. The maximum charge swing across the tank capacitance can easily be expressed in terms of the frequency of oscillation, the voltage swing and the tank inductance as

$$q_{max} = C_{tank}V_{sw} = \frac{V_{sw}}{\omega_o^2 L_{tank}} \tag{2.18}$$

where V_{sw} denotes the voltage swing across the LC tank, L_{tank} and ω_o denote the tank inductance and the frequency of oscillation, respectively. Substituting (2.18) into (2.17), the spectrum in the $1/f^2$ region is obtained as

$$\mathcal{L}\{\Delta\omega\} = \frac{\omega_o^4}{2\Delta\omega^2} \cdot \frac{L_{tank}^2}{V_{sw}^2} \cdot \sum_n \left(\frac{\overline{i_n^2}}{\Delta f} \cdot \Gamma_{rms,n}^2 \right) . \tag{2.19}$$

As the impulse sensitivity function is periodic, it can be expressed as a Fourier Series

$$\Gamma(\omega_o\tau) = \frac{c_0}{2} + \sum_{n=1}^{\infty} c_n \cos(n\omega_o\tau) \tag{2.20}$$

where the coefficients c_n are real. c_0, the dc coefficient of ISF governs the upconversion of $1/f$ device noise around the carrier, whereas the other coefficients are responsible from the downconversion of white noise near the harmonics of the carrier, [30], [40].

The spectrum of phase noise in the $1/f^3$ region, where the upconversion of device $1/f$ noise around the carrier is observed is described by, [41]

$$\mathcal{L}\{\Delta\omega\} = \sum_n \frac{\frac{\overline{i_n^2}}{\Delta f} c_0^2}{8q_{max}^2 \Delta\omega^2} \cdot \frac{\omega_{1/f}}{\Delta\omega} \tag{2.21}$$

where $\omega_{1/f}$ denotes the $1/f$ noise corner of the associated device in the oscillator. As the equations describing the spectrum of phase noise show, the ISF of all the noise-generating components should somehow be calculated to compute their contribution to the phase noise of the VCO. Once the individual contributions are computed, the phase noise of the VCO can be determined through summation as (2.17) and (2.21) suggest. Note that each $\overline{i_n^2}/\Delta f$ in the equations represents drain current noise, resistor noise, varactor noise and inductor noise. The varactor and inductor noise are due to the finite quality factor, Q, of the components and they can easily be expressed in terms of the loss resistances of the components which can be obtained from their respective simulated Q values at the frequency of oscillation. The mean-square thermal noise current density of a resistor and the mean-square drain current thermal noise density of an MOS transistor are given as, [72]

$$\begin{aligned} \frac{\overline{i_R^2}}{\Delta f} &= \frac{4kT}{R} \\ \frac{\overline{i_d^2}}{\Delta f} &= \frac{8kT}{3} \cdot (g_m + g_{mbs} + g_{ds}) \end{aligned} \tag{2.22}$$

where g_m, g_{mbs} and g_{ds} denote the transconductance gain, the bulk transconductance and the channel conductance of an MOS transistor, respectively. The mean-square flicker noise current density of an MOS transistor,on the other hand, is given as

$$\frac{\overline{i_d^2}}{\Delta f} = \frac{K_f I_{ds}^{a_f}}{C_{ox} L_{eff}^2 f^{e_f}} \tag{2.23}$$

where I_{ds} is the dc drain current, C_{ox} is the gate oxide unit capacitance, L_{eff} is the effective channel length of the transistor, K_f is the flicker noise parameter, a_f is the flicker exponent and e_f is the frequency exponent, [72].

To determine the ISF of the noise-generating components of the VCO, the noise sources in the circuit were replaced with impulsive current sources of small area one by one and the resulting phase shift was measured through extensive Spectre transient simulations through Cadence, as explained in [30]. The area of the current impulse injected corresponds to the amount of charge perturbation applied to the circuit and in order for linearity, which is one of the bases of the theory, to hold, the amount of charge injected should be much smaller than the steady-state charge swing across the tank capacitance. A scaling on the measured phase shift was done after the transient simulations, accordingly. The schematic shown in Fig.27 depicts each component with its associated noise source as used in the phase noise analysis. Note that, for the sake of clarity in the schematic, the thermal noise due to the resistive loss of the LC tank is lumped into a single noise source denoted as $\overline{i_{tank}^2}$.

The procedure roughly explained above for determining the ISF was repeated for each component with a noise source shown in Fig.27, where $R_{s3} = R_{s4} = 0$ was used in the circuit designed. Due to the symmetry in the fully-differential circuit, one set of transient simulations were run for the identical components. As mentioned before, the value of ISF, $\Gamma(\omega_o\tau)$, depends on at what phase, $\omega_o\tau$, the current impulse is injected into the circuit and it is this characteristic of the ISF that accounts for the time-variance involved in the theory. In order to determine the Fourier coefficients of the ISF of a noise-generating component accurately, the number of current impulse

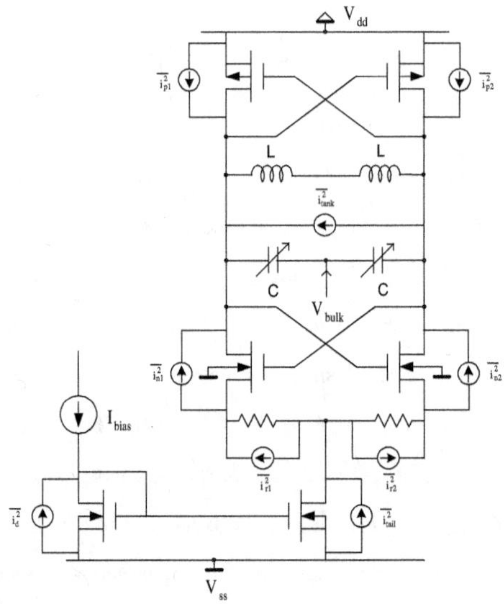

Fig. 27. CMOS voltage-controlled oscillator with noise sources for phase noise analysis.

injection simulations within an oscillation period, which corresponds to 2π radians, should be kept as high as possible. The current impulse is injected once the oscillating waveform reaches its steady-state. The simulations are repeated N times by changing the phase at what the current impulse is injected within one oscillation period, i.e. 2π radians, with $2\pi/N$ increments each time with respect to the starting phase. Obviously, the accuracy with which (2.17) and (2.21) estimate the phase noise depends on the accuracy with which the Fourier coefficients are determined. As an approximate figure for the phase noise of the VCO was targeted for the analytical calculations, the mean-square value of the ISFs of the tank inductor and the tank varactor were taken as 1/2, as is the case for an ideal sinusoidal waveform of an LC oscillator, [30]. Furthermore, the number of transient simulations for the ISF determination of each component was kept at a relatively small number of 50. The simulated ISFs of the NMOS and PMOS cross-coupled transistors and that of the tail transistor are plotted in Fig.28. The mean-square value of the ISF of the PMOS transistor is obtained as 0.27 and it is higher than that of the NMOS transistor, which is 0.13. The reason for the difference is anticipated to be due to the fact that half-circuit symmetry properties ($g_{mn} = g_{mp}$) were not observed in the design phase. The dc value of the ISF of the NMOS cross-coupled transistor was extracted as 0.033, whereas that of the PMOS cross-coupled transistor was extracted as -0.033. It should be remembered that the dc value of the ISF governs the upconversion of $1/f$ device noise around the carrier and the factors with which both cross-coupled transistors contribute to the upconversion of flicker noise around the carrier are the same. As the flicker noise of the NMOS device is much higher than that of the PMOS device in the technology used, the $1/f$ noise contribution to the oscillator phase noise from the NMOS transistor is much higher. On the other hand, as it can be seen in Fig.28, the mean-square value of the ISF of the tail transistor, which is obtained as 0.015, is much smaller than those of the cross-coupled transistors. Thus, the upconverted white-noise contribution from the tail transistor is considerably less than the contributions of the cross-coupled transistors. The dc value of the ISF of the tail transistor is extracted as -0.09, a value higher than those of the cross-coupled transistors. Although the effective channel length of the tail transistor is twice the channel lengths of the cross-coupled transistors, which are with minimum effective channel length of $0.5\mu m$ available in the technology, the higher ISF dc value causes the upconverted $1/f$ noise contribution of the tail transistor to be comparable to the contributions of the cross-coupled transistors (note that there are total of four cross- coupled transistors generating noise, whereas there is only one tail transistor). A noteworthy observation from Fig.28 is that two periods of the ISF of the tail transistor exist within one period of the ISFs of the cross-coupled transistors. The reason for this phenomenon is the fact that the drain of the tail transistor swings at twice the frequency of the oscillation as it is pulled-up each time one of the cross-coupled NMOS transistors is turned on, [30]. It should be pointed out that smaller ISF dc values could have been obtained for the cross-coupled transistors by observing half-circuit symmetry properties through the proper sizing of the NMOS and PMOS transistors for equal transconductance gain. Such an optimization at the design phase is required to reduce close-in phase noise. The flicker noise upconversion of the tail transistor is also a significant contributor to the phase noise and the optimization through exploiting the symmetry properties mentioned above also reduces the contribution from the tail transistor. Furthermore, cascoding techniques at the tail current source and a relatively large capacitance at the common-mode node of the tail current source may help reduce the sensitivity of the oscillator output to the device flicker noise of the cross-coupled transistors through reducing the dc value of their ISFs, and the white noise of the tail transistor through reducing the mean-square value of its ISF, [40], [30].

The simulated ISFs of the diode-connected current source transistor and the source degeneration resistors are shown in Fig.29. Note that the ISF of the diode-connected transistor in the tail current source swings twice within one oscillation period as the ISF of the tail transistor does for the same reason. The rather sharp ISF of the source-degeneration resistor is due to the fact that less simulations were performed to extract its ISF. Increasing the number of charge injection simulations would smooth out the ISF as the ISF of the diode-connected transistor in the same figure obtained with more simulations shows. The dc and mean-square values of the ISF of the diode-connected transistor are computed as 0.8 and 0.65, respectively. Those figures indicate that the flicker noise and thermal noise contributions of this transistor is very pronounced at the output phase noise. As it is seen from Fig.29,

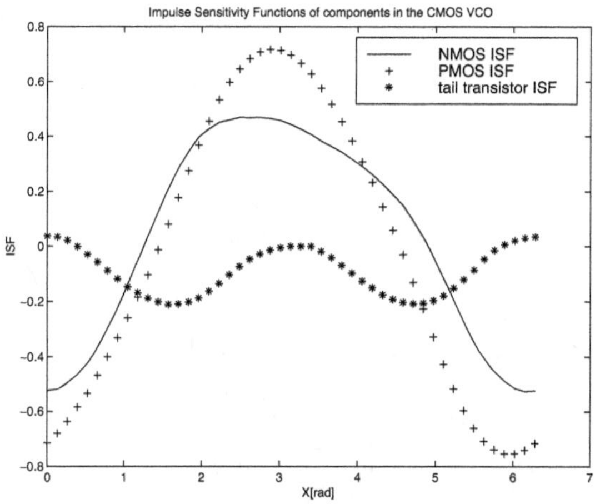

Fig. 28. Simulated ISFs of the cross-coupled transistors and the tail transistor in the CMOS VCO.

the mean-square value of the ISF of the source-degeneration resistors is much lower with a value of 0.019. The source-degeneration resistors were implemented with the only poly layer of the technology and $1/f$ noise of the resistor was neglected as no related model parameters were available.

The transient charge injection simulations and the nonlinear phase noise simulations were performed with Spectre and SpectreRF in Cadence, respectively. The nonlinear phase noise analysis of SpectreRF called as Periodic-Steady State simulation provides the phase noise contributions of the components in the VCO and the results obtained from the simulations agree reasonably with the contribution figures calculated analytically using (2.17) and (2.21). A comparison between the analytically calculated and the SpectreRF simulated phase noise of the CMOS VCO core is plotted in Fig.30, where the agreement between the two is quite satisfactory.

The analytical calculations were also compared with the measured results as well

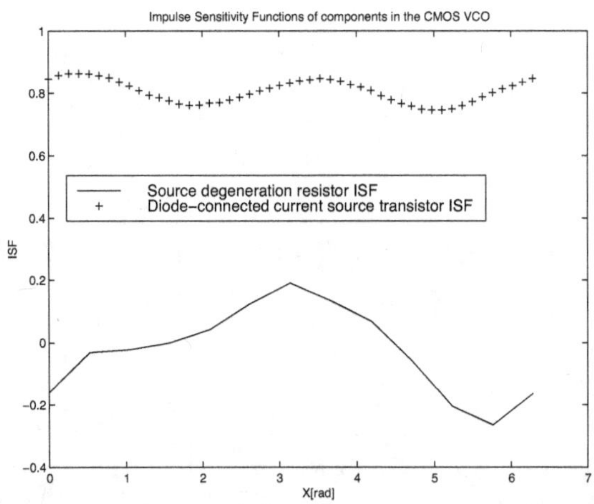

Fig. 29. Simulated ISFs of the source-degeneration resistors and the diode-connected current source transistor in the CMOS VCO.

as the results of the SpectreRF simulations. As the analytical results were to be compared to the measured results, the output buffer, the somehow simplified package model, the passive by-pass components on the PCB (used for the supply and dc bias lines) were all added to the test bench used in the charge injection simulations. A significant decrease was observed in both the dc and the mean-square values of the ISF of the diode-connected transistor in the tail current source, mainly due to the addition of relatively high capacitances connected to the drain node of the transistor (in the order of nF) representing the surface mount capacitors soldered onto the off-chip bias line on the PCB. As the diode-connected transistor was a significant contributor to the phase noise of the CMOS VCO core, especially at low offset frequencies, a considerable improvement in the close-in phase noise was observed in both the simulations and the analytical calculations.

The calculated contributions of the components in the VCO to the output phase

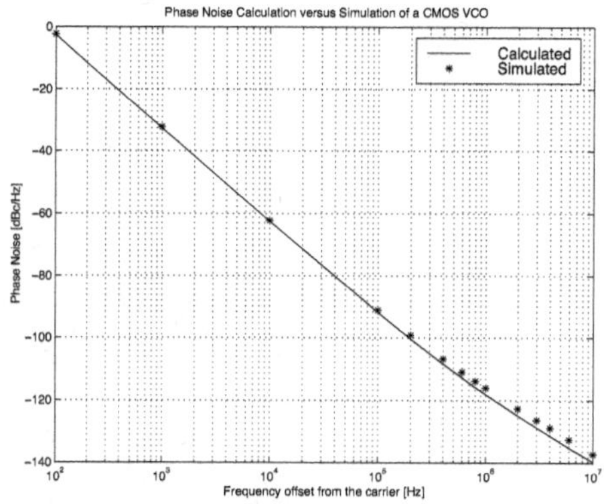

Fig. 30. Calculated (with Eqs. (2.17) and (2.21)) vs. SpectreRF simulated phase noise of the CMOS VCO.

Table II. Calculated phase noise percentage contributions of the VCO components 1.

	NMOS pair	PMOS pair	R_S	Inductors
1kHz	50%	5.3%	0.04%	0.4%
600kHz	20.6%	12.3%	4.2%	50.7%
3MHz	15%	14%	5%	60%

noise at 1 kHz, 600kHz and 3 MHz offsets from the carrier are tabulated in Tables II and III, with the varactors operating in depletion. As the percentage contributions at 600kHz and 3MHz offsets clearly show, the low Q inductor is the main contributor to the phase noise at large offset frequencies. Note that at low offsets from the carrier, i.e. at 1kHz offset shown in the tables, with the addition of the upconverted flicker noise of the transistors, the percentage contributions differ considerably. The largest contributors are the NMOS cross-coupled pair and the tail transistor, while the PMOS cross-coupled pair contributes much less with its superior flicker noise

Table III. Calculated phase noise percentage contributions of the VCO components 2.

	Tail transistor	Diode transistor	Varactors
1kHz	33%	12%	0.01%
600kHz	8.3%	2.7%	1.2%
3MHz	4%	1%	1.4%

in the technology. By observing the symmetry properties of the oscillating waveform through the proper sizing of the cross- coupled transistors and the source-degeneration resistors, it is possible to reduce the dc values of the ISFs of the MOS transistors, which govern the upconversion of the flicker noise around the carrier, to improve the close-in phase noise, [30]. As for the phase noise at large offsets, an integrated LC tank with higher quality factor should be used, as the calculated percentage contributions at 600kHz and 3 MHz offsets from the carrier given above clearly show. Since the Q of the tank is determined by that of the inductor, to demonstrate how the Q of the inductor affects the phase noise of the VCO at large offsets quantitatively, SpectreRF simulations were run using the VCO test bench, by setting the Q of the tank inductor as a parameter, keeping the bias current constant. The simulation results are plotted in Fig.31, where the y-axis is the phase noise at 3 MHz offset in dBc/Hz and the x-axis is the Q of the inductor. The improvement in the phase noise as the inductor Q increases, especially up to a value of 15, as observed in Fig.31 clearly illustrates the dominant contribution of the inductor noise in the phase noise of the VCO at large offsets. At inductor Q values higher than 15, the improvement becomes marginal as the contributions from the cross-coupled pairs, the tail transistor and the source-degeneration resistors start to dominate. The pie charts in Fig.31 show the relative phase noise contributions of the tank inductor and the other noise-generating components for Q values of 2.5 (as used in the design), 5 and 10. The graphical illustration demonstrates that as the Q of the tank inductor increases, its phase noise contribution decreases and the other noise-generating components start to dominate the phase noise of the VCO.

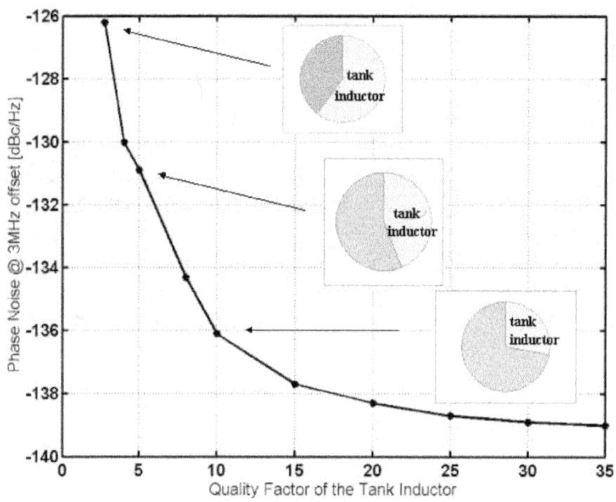

Fig. 31. Simulated phase noise of the CMOS VCO @ 3MHz offset vs. the quality factor of the tank inductor.

The source-degeneration resistors are used to reduce device noise upconversion through linearization. In the CMOS oscillators, the reduction is mainly for the flicker noise upconversion. The addition of the degeneration resistors makes the dc values of the ISFs of the cross-coupled transistors and the tail current source transistors lower thus reducing their contributions in the $1/f^3$spectrum of the phase noise.

The trade-offs involved in using the source-degeneration resistors from the phase noise point of view were inspected through SpectreRF simulations of a test bench, which includes the VCO core, the output buffer, a simplified package model and the

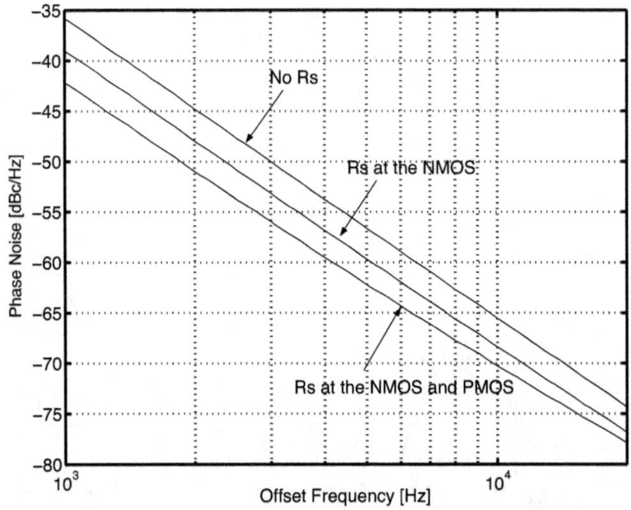

Fig. 32. Simulated effect of the source degeneration resistors on the close-in phase noise of the CMOS VCO.

by-pass components of the supply lines on the PCB. Figs.32 and 33 compare the phase noise at low offset frequencies and large offset frequencies from the carrier, respectively. The three curves in each plot correspond to the cases: i- without any degeneration resistor (No R_S), ii- with R_S only at the NMOS and iii- with R_S at both the NMOS and PMOS cross-coupled transistors. Although a significant improvement is observed at close-in phase noise in Fig.32 due to the reduction of the upconverted flicker noise of the transistors with the addition of the resistors; as the offset from the carrier gets larger, the contribution from the upconverted thermal noise of the resistors becomes more and more pronounced deteriorating the phase noise of the VCO, as clearly seen in Fig.33. In other words, *the source-degeneration resistors trade the close-in phase noise with the phase noise at large offsets.* The $1/f^3$ corner

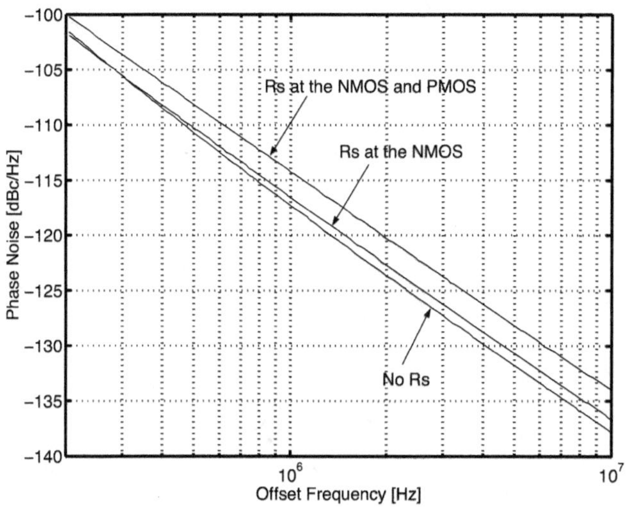

Fig. 33. Simulated effect of the source degeneration resistors on the phase noise at large offsets from the carrier of the CMOS VCO.

frequencies get smaller with the addition of the source-degeneration resistors. If no degeneration is used in the cross-coupled pairs, the simulated $1/f^3$ corner frequency is obtained at around 250kHz, whereas the resistors at the NMOS cross-coupled pair reduce the $1/f^3$ corner frequency to around 60kHz. When both cross-coupled pairs are degenerated, the resulting $1/f^3$ corner frequency is simulated to be around 4kHz. If the specifications of the VCO to be designed requires low phase noise at low offsets from the carrier, degeneration resistors can be added either only to the NMOS cross-coupled pair or to both the NMOS and PMOS cross-coupled pairs depending on the available margin at the phase noise specifications at large offsets. The effect of the upconverted thermal noise contributed by the resistors is observed at large offsets from the carrier where the advantage of using the source-degeneration resistors for phase noise concerns seems to be no longer pronounced. An additional factor in the increase of the phase noise at large offsets is that with the linearizing degeneration resistors, the effective transconductances of the cross-coupled pairs are reduced decreasing the oscillation amplitude as shown in the simulation results given in Fig.34. The reduced oscillation amplitude increases the phase noise contribution of the components in the $1/f^2$ region, i.e. at large offsets, as the term q_{max} in (2.17) indicate.

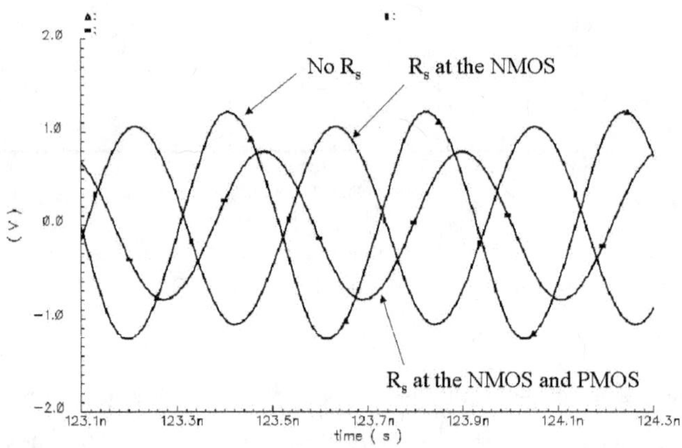

Fig. 34. Simulated effect of the source degeneration resistors on the oscillation amplitude of the CMOS VCO.

The linearizing effect of the source-degeneration resistors is observed in the simulations where the magnitude of the third-order harmonic relative to the fundamental is decreased with the addition of the resistors. When no degeneration is used, the third-order harmonic is 35dB below the fundamental, whereas with degeneration at the NMOS pair, the difference between the fundamental and the third-order component becomes 40dB. In the case of degenerating both cross-coupled pairs, the third-order harmonic is simulated to be around 47dB below the fundamental. On the other hand, it can be anticipated that, the addition of the resistors R_{s3} and R_{s4} to the PMOS transistors would improve the power supply rejection ratio (PSRR), i.e. the pushing figure of the oscillator. Note that the source degeneration resistors reduce the loop gain for a given bias current. Therefore, they could also reduce the tuning range for a given VCO bias current, in case the Q of the varactor lowers the overall tank Q within the capacitance tuning range. This is observed quite clearly in the comparison of the measured frequency tuning ranges of the three VCOs presented in Chapter V.

CHAPTER III

DUAL-MODULUS PRESCALER DESIGN IN BICMOS

Prescalers are among the key building blocks of GHz-range frequency synthesizers used in wireless communication systems, [73], [74]. The quality of the prescaler is vital in setting the performance of the PLL-based frequency synthesizers used as Local Oscillators, especially from the power consumption point of view. There is a continuing research effort to improve the efficiency (i.e. higher operating frequency with lower power consumption) of the prescalers designed with the mainstream CMOS and BiCMOS technologies, [42], [33]. However, the impact of the prescaler phase noise on the PLL output phase noise has not received much attention in the literature. The noise contributors in the prescalers are hardly investigated in the papers causing a lack of understanding to optimize the noise performance of the prescalers designed for wireless communication systems. In this chapter, after briefly emphasizing the impact of the prescaler noise on the phase noise performance of the PLL-based frequency synthesizers in Section A, a dual-modulus prescaler design in a BiCMOS technology is presented in Section B. The trade-offs involved in the design are mentioned with emphasis on the phase noise performance. Section C is a literature survey on the injection-locked frequency dividers. The measurement results of the prescaler are provided in Chapter V.

A. Prescalers in PLLs

PLL-based frequency synthesizers are widely used in modern communication systems, mostly in synthesizing well defined, stable local oscillator signals.

One of the most challenging parts of especially high-frequency PLLs used in wireless communication systems is the prescaler due to the demanding specification

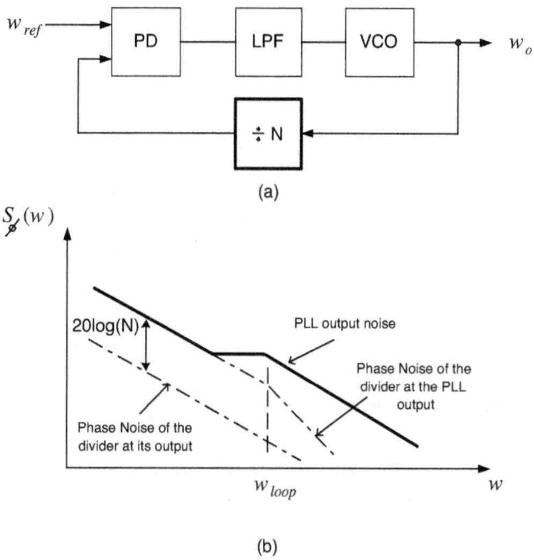

Fig. 35. Phase-locked loop: (a) basic block diagram, (b) impact of the divider phase noise on the phase noise of a first-order loop with a low noise input.

of low power consumption and low phase noise at GHz frequencies. Block diagram of a basic PLL is shown in Fig.35(a), where the fixed mode prescaler is embedded into the divider in the feedback loop for the sake of simplicity. The phase noise of a first-order PLL with a low noise input is depicted in Fig.35(b) for the sake of illustrating the effect of the divider phase noise on the output noise of the overall PLL system, [30]. As it is shown in the figure, the phase noise at the output of the divider is multiplied by the square of the divide ratio N when translated to the PLL output phase noise. Note that, even though the divider phase noise at offset frequencies larger than the loop bandwidth is suppressed by the loop, the divider phase noise at offsets lower than the loop bandwidth, denoted as ω_{loop}, contributes to the overall phase noise of the PLL. In the state-of-the-art fractional-N frequency synthesizers, the loop bandwidth values are usually at a few tens of kHz and most of the wireless standards have demanding phase noise specifications at offsets lower than a few tens of kHz (e.g., DCS 1800 specification of -80dBc/Hz at 1kHz). This clearly illustrates the need for optimizing the prescalers for low noise.

B. Dual-Modulus BiCMOS Prescaler Design

The block diagram of the dual-modulus prescaler depicted for single-ended processing for the sake of clarity is shown in Fig.36. The input buffer functions as a single-ended to differential converter driving the synchronous divide by 4/5 counter. The output of the synchronous counter drives the divide by 8 extender which is comprised of three consecutive asynchronous divide by 2 stages. The additional logic in the feedback loop between the outputs of the asynchronous dividers and the synchronous divide by 4/5 counter is designed to modify the divide ratio in accordance with the $Mode$ signal. Differential ECL-based D flip-flop (DFF) and NOR gates are adopted to improve the immunity to common-mode noise sources such as the noise on the substrate and on the power supply lines.

The use of ECL level signals in the digital building blocks prove advantageous over the use of CMOS levels in the integrated circuits which accommodate sensitive analog circuits on the same die from the point of view of reducing the disturbance to the sensitive analog signal lines. In the context of the PLL-based frequency synthesizer design, an example of such a sensitive analog signal line is the frequency control input line of the VCO. The switching noise coupled from the prescaler would disturb the VCO output significantly, creating spurs at the synthesizer output. The ECL-level signals would create less noisy disturbances on the substrate than the CMOS-level signals, providing more immunity to the VCO frequency control input.

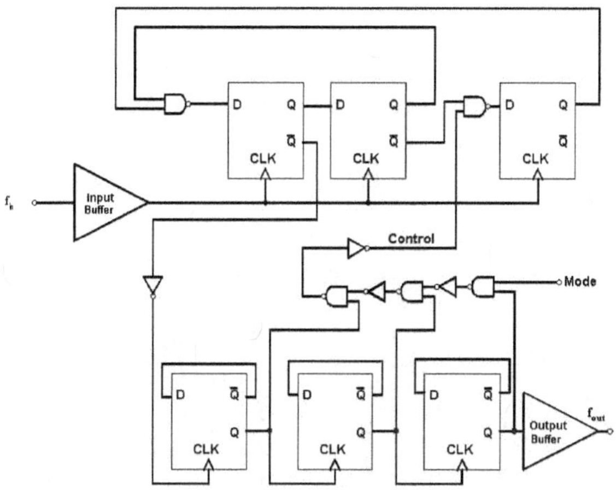

Fig. 36. Block diagram of the dual-modulus prescaler.

The biasing of the prescaler is supplied by a bandgap bias generator from which all the bias currents of the building blocks are distributed. The phase noise contribution of the bias generator can be significant, requiring proper measures to be taken to design the circuit for low noise, as it will be clear in the following. The integrated bandgap bias network that provides the bias currents of the building blocks of the prescaler is shown in Fig.37.

The prescaler was designed to process a single-ended signal coming from an off-chip VCO. The input buffer acts like a single-ended to differential converter for this purpose. The output of the input buffer provides the ECL level differential clock signals to the synchronous divide by 4/5 stage. The buffer is a simple differential pair followed by emitter followers to drive the resistive-capacitive loads in the synchronous divide-by-4/5 stage. Note that the buffer should be designed carefully to assure a reasonable input sensitivity with as low of a power consumption as possible. The

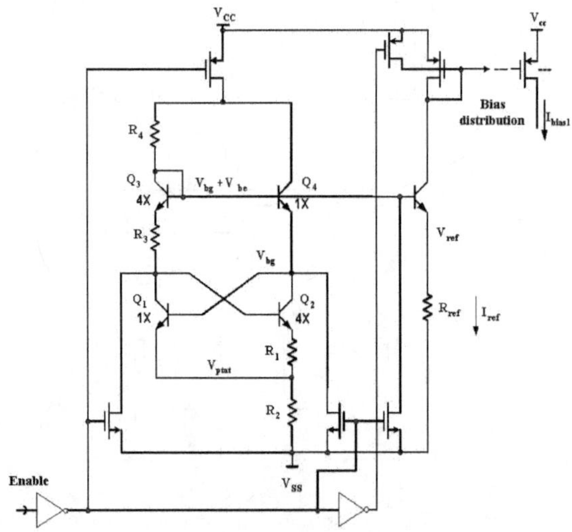

Fig. 37. Bandgap bias generator of the prescaler.

schematic of the buffer is given in Fig.38.

The DFFs and the NOR gates used in the dividers are all ECL-based building blocks. The schematic of the master-slave DFF is shown in Fig.39. Scaling of the bias currents were done in the dividers in accordance with the frequency of the signals they have to process. The emitter followers were omitted wherever possible to reduce the power consumption. All the gates use minimum emitter size transistors on their signal path to reduce the capacitive loading and thus to minimize the power consumption. To comply with the next generation portable devices, the power supply voltage was tried to be minimized. NMOS transistors were preferred over the bipolar transistors as the tail bias transistors of the gates to reduce the required voltage headroom for minimizing the supply voltage. The circuit that limits the reduction at the power supply voltage is the bandgap generator and 2.5V is used. The NOR gate which

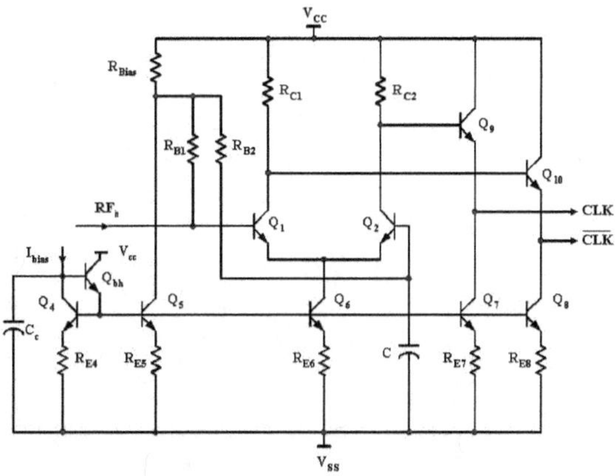

Fig. 38. The input buffer.

has the off-chip CMOS level Mode signal at one of its inputs is a true BiCMOS gate that works with CMOS levels at one input and ECL levels at the other input. The schematic of the gate is shown in Fig.40.

Most important design trade-off in the prescalers is the one between the speed of operation and the power consumption. Special care has been taken in the design to reduce the power consumption of the blocks while guaranteeing proper operation at 2.3GHz. The simulated power consumption contributions of the different sections of the dual modulus prescaler are tabulated in Table IV. The main contributors are the building blocks which run at the maximum input frequency, i.e., the synchronous divide by 4/5 counter and the input buffer. The floorplan of the layout was carefully designed in order to minimize the interconnection capacitances in the critical paths.

The phase noise of the prescaler was investigated through simulations to determine the main noise contributors. The simulated phase noise of the prescaler obtained from Periodic Steady State simulations in SpectreRF of Cadence is shown in Fig.41.

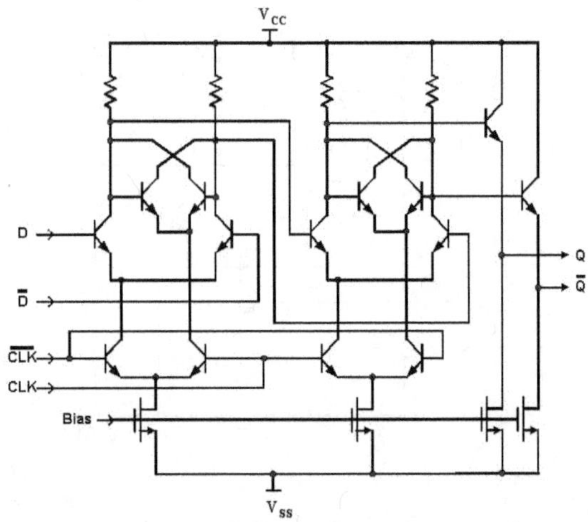

Fig. 39. The ECL-based master and slave DFF.

Table IV. Simulated power consumption percentage contributions of the dual-modulus prescaler building blocks.

Input Buffer	Bias Network	Synchronous Counter	Asynchronous Dividers
24%	8.3%	48.3%	19.2%

As the PLL system the prescaler is aimed to be used in will shape the noise contributed by the prescaler with a lowpass characteristic, the corner frequency of which is determined by the loop bandwidth, the phase noise contribution from the prescaler at low offset frequencies (i.e. the frequencies lower than the loop bandwidth which is usually at a few tens of kHz in the frequency synthesizers used in the wireless communication systems) are of importance.

The main contributors to the phase noise at low offsets are found to be the

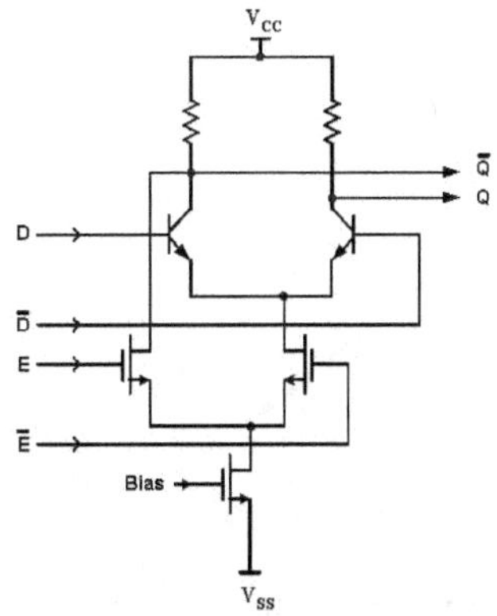

Fig. 40. The true BiCMOS NOR gate with ECL and CMOS level inputs.

1/f noise of the MOS current mirror transistors providing the bias to the divide by 8 extender asynchronous dividers. The diode-connected NMOS tail current mirror transistor biasing the asynchronous dividers contributes almost 62% of the total noise at 1kHz and 5kHz offsets. The bias currents of the building blocks are distributed from the bandgap bias generator using PMOS current mirror transistors and the 1/f noise of the PMOS transistors providing the bias to the tail current mirror of the extender asynchronous dividers accounts for about 12% of the total phase noise at both 1kHz and 5kHz offsets. A straightforward way to reduce the 1/f noise of the MOS transistors is to increase the gate lengths and simply by doubling the gate widths and lengths of the MOS transistors, a reduction of about 6dB is obtained in the phase noise at 1kHz offset as shown in the simulation results given in Fig.41. A better solution to reduce the noise would be to use bipolar transistors at the current mirrors for biasing the ECL gates, as the bipolar devices generate much less 1/f noise than the MOS devices, but this would conflict with the reduction of the supply voltage

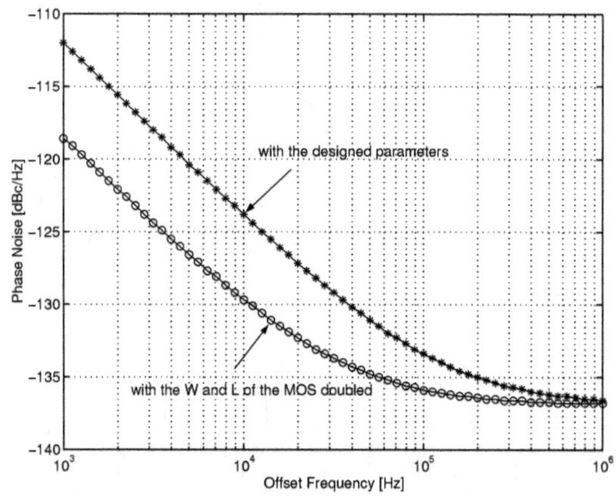

Fig. 41. Simulated phase noise of the prescaler.

and hence the power consumption as mentioned earlier. The trade-off between phase noise and power consumption is obvious here.

Note that in Fig.41, no phase noise improvement is observed at the noise floor reached at higher offsets by doubling the gate areas, because an improvement at the noise floor requires a reduction at the thermal noise contributions of the devices by increasing the current consumption, imposing a trade-off between power consumption and the noise floor, [42], [33]. As the PLL already suppresses the phase noise of the prescaler at high offset frequencies, it is not advisable to burn more power to reduce the noise floor of the prescaler itself.

C. Injection-Locked Frequency Dividers

High-speed frequency dividers, which are needed in optical and satellite communication systems, are often used as indicators for new high-speed technologies, [75]. The main use of the frequency dividers in the communication systems are in the phase locked loop frequency synthesizers to divide the VCO output frequency to the values low enough that can be processed with programmable dividers in the feedback loop, [7]. The development in broadband networks such as the 10Gb/s ethernet increases the demand of realizing the required circuits including the frequency dividers in low cost mainstream digital CMOS technologies. The existing high-speed frequency-division techniques can be categorized into two groups: wide-band flip-flop based, [76], [75], [42] and narrow-band resonator based (injection-locked dividers), [77], [71]. The wide-band flip-flop based division technique has the drawback of the strong trade-off it involves between the operating speed and power consumption. These structures need quite high current consumptions to reach operating speeds exceeding 10GHz required by the emerging technologies. Therefore, in the narrow band applications such as the wireless communication systems, the resonator-based frequency dividers can be employed to reduce the power consumption and increase the maximum operating frequency, [78]. The main reason why the wide-band flip-flop based dividers suffer from high power consumption is due to the complete charging and discharging of capacitances during each cycle. In the contrary, the injection-locked dividers dissipate a fraction of the energy stored in the tank, which is determined by the quality factor Q of the tank, in every cycle, [71]. It should be noted that the dynamic loading in the flip-flop based dividers as proposed in [75] helps to reduce the power consumption but the injection-locked dividers still provide solutions with lower power consumption.

Injection locking can be used to frequency lock an oscillator to an external signal called as incident signal, [79]. When the frequency of the incident signal is a harmonic of the oscillator signal, then frequency division is performed, [80]. The oscillation conditions of unity loop gain and zero excess phase have to be satisfied for the locking to occur in the injection-locked frequency dividers (ILFDs). The locking range of an ILFD can be limited by failure of either the phase or the gain condition. The model of an injection-locked frequency divider (ILFD) is shown in Fig.42, [80].

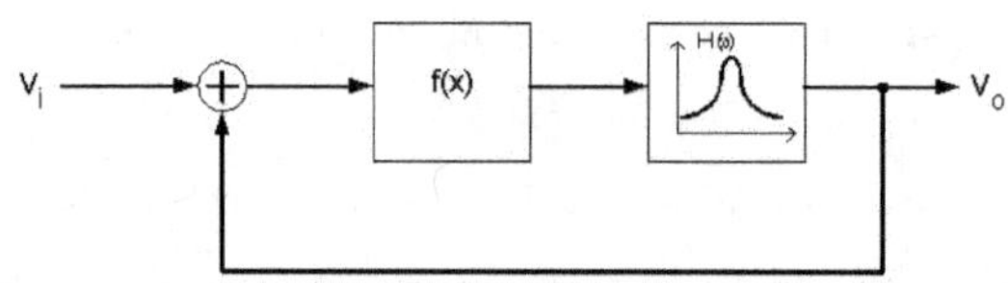

Fig. 42. Model for an injection-locked frequency divider.

The function $f(x)$ models all the nonlinearities in an ILFD and $H(\omega)$ represents the frequency selective block. In the special case of a third-order nonlinearity at $f(x)$ and a divide-by-two operation, the phase-limited locking range of an LC ILFD can be expressed as:

$$\left|\frac{\Delta\omega}{\omega_r}\right| < \left|\frac{H_0 a_2 V_i}{2Q}\right| \tag{3.1}$$

where ω_r is the free-running oscillation frequency, $\Delta\omega$ is the frequency offset from ω_r, V_i is the incident amplitude, H_0 is the impedance of the LC tank at resonance, Q is the quality factor of the LC tank at resonance, and a_2 is the second-order coefficient of the nonlinearity in $f(x)$, [80]. Note that a larger H_0/Q results in a larger locking range for a given incident amplitude. In an LC oscillator $H_0/Q = \omega L$, so using the largest practical inductance would help to maximize the locking range, [80]. Also note that the locking range increases as the amplitude of the incident signal V_i increases, which is intuitive.

1. Circuit Implementations

The schematic of a single-ended injection-locked frequency divider based on a Colpitts oscillator core is shown in Fig.43, [78]. The incident signal to be divided by two in frequency is injected into the gate of transistor M_1 through the AC coupling capacitor C.

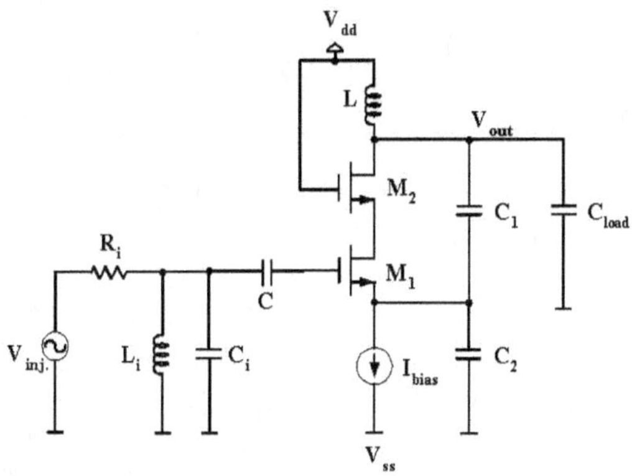

Fig. 43. Single-ended injection-locked frequency divider.

The parallel resonator circuit of L_i and C_i are representing the tank of the LC VCO driving the divider. The bias circuit of the Colpitts oscillator is not shown for simplicity. Transistor M_2 is added to improve the isolation between the output and the input of the divider. In the design of the cascode transistors, it is advisable to size transistor M_2 to be smaller than M_1 to reduce the parasitic capacitance at the output node. The reduction at the parasitic capacitance is important because in this way, a larger inductor can be used to resonate the capacitance. A larger inductor would help to increase the locking range by increasing the ratio of the impedance of the LC tank to its quality factor at resonance, i.e. H_0/Q as suggested by Eq.3.1, [78]. Additionally, the power consumption of the core would also decrease due to the increased effective parallel impedance of the tank with a larger inductor, assuming that the loss in the tank is dominated by that of the inductor. The integrated passive elements L, C_1 and C_2 form the tank circuit of the Colpitts oscillator core, the frequency of which is locked to half the frequency of the incident signal by proper design of its tank elements. C_{load} denotes the load capacitance and it can be the input capacitance of the dual-modulus prescaler in the feedback of the phase locked loop within which the divider is used, [81]. An analogy of this circuit with the model of the injection-locked divider depicted in Fig.42 suggests that transistor M_1 functions as the summing element for the incident and the output signals. The circuit was implemented with a $0.5\mu m$ CMOS technology and operates with a supply voltage of 2.5V drawing 1.2mA. The free-running frequency is 920MHz and the injection (incident) frequency is 1840MHz. The total die area is 0.7 mm^2. A locking range more than 190MHz is measured with a power consumption of 3mW.

Schematic of the differential injection-locked divider based on a negative conductance LC oscillator is shown in Fig.44, [78]. The injection signal is applied to the gate of the tail transistor M_3 through the AC coupling capacitor C. This transistor delivers the injected power to the common source node of the cross-coupled differential pair. The output signal is fed back through the cross coupling. Once again with the analogy to the model shown in Fig.42, the incident signal and the output signal are summed across the gates and sources of transistors M_1 and M_2, [78]. Note that the common source node of the cross-coupled differential pair oscillates at twice the frequency of the output signal even in the absence of the incident signal, which makes this node very appropriate to be used as the injection node for a divide-by-two operation. The differential ILFD was designed and implemented with a $0.5\mu m$ CMOS technology.

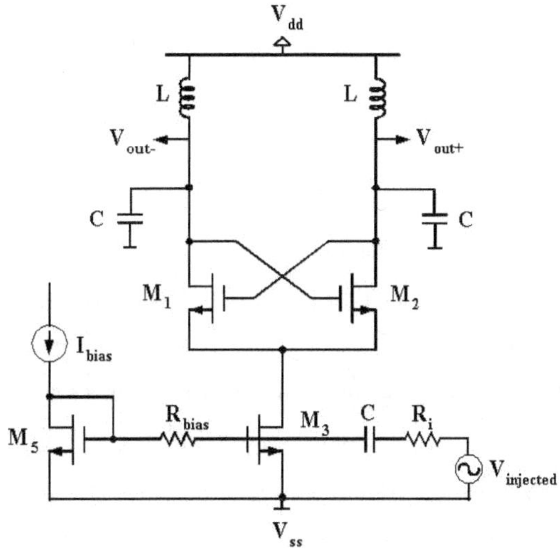

Fig. 44. Differential injection-locked frequency divider (DILFD).

The circuit operates with a supply voltage of 1.5V drawing 300μA. The free-running oscillation of the DILFD is at 1.6GHz while the incident signal is at 3.2GHz. The measured results presented in the paper show more than 190MHz locking range with only 0.45mW of power. When the power is increased to 1.2mW, the locking range increases to 370MHz, which is 12% of the center frequency. For comparison purposes, a digital frequency divider using a Source-Coupled-Logic (SCL) master-slave D flip-flop connected in divide-by-two configuration was also implemented and measured. It is reported that the SCL divider operates at about half the frequency of the DILFD and consumes more than four times power.

As it was mentioned before, locking range in the injection-locked dividers is limited. It is desirable to increase the locking range of the structures but still keeping their low power aspect. Two architectural solutions were proposed to overcome this limitation. The first one presented in [80] incorporates varactors into the differential injection-locked divider of Fig.44 as tank elements to enable the free-running oscillation frequency of the divider to track the frequency of the VCO generating the divider incident signals in a PLL application. A 29% measured locking range is reported for an integrated implementation with a $0.24\mu m$ CMOS technology. The IC measurements show frequency division at 5GHz with more than 1GHz locking range and power consumption of less than 1mW, [80]. Successful implementation of the circuit in a fully-integrated CMOS frequency synthesizer for a 5GHz Wireless Local Area Network (LAN) receiver is presented in [81] and [82]. The divider is a replica of the VCO and with the use of varactors driven by the control voltage of the VCO, it tracks the output frequency of PLL. The output of the divider drives the dual-modulus prescaler. The synthesizer was implemented with a $0.24\mu m$ CMOS technology and the voltage controlled injection-locked divider dissipates 1.2mW power from a 1.5V supply providing a locking range that encloses the tuning range of the VCO as expected.

The other solution presented in [71] is to use shunt-peaking to increase the locking range and lower the power dissipation at higher frequencies. The differential injection-locked divider is used as the core of the circuit as shown in Fig.45.
The total parasitic capacitance at the common-source node of the cross-coupled differential pair is explicitly represented by the capacitor C_{tail} in the schematic. The impact of this capacitance is to significantly lower the effective internal injection power that otherwise can be used for injection locking, [71]. To remedy this problem, a shunt inductor L_{shunt}, is added to the circuit to resonate with C_{tail} at the injection frequency as shown in Fig.45. In this way, the impedance of the tail node at the injection frequency increases and so does the injection power transferred to this in-

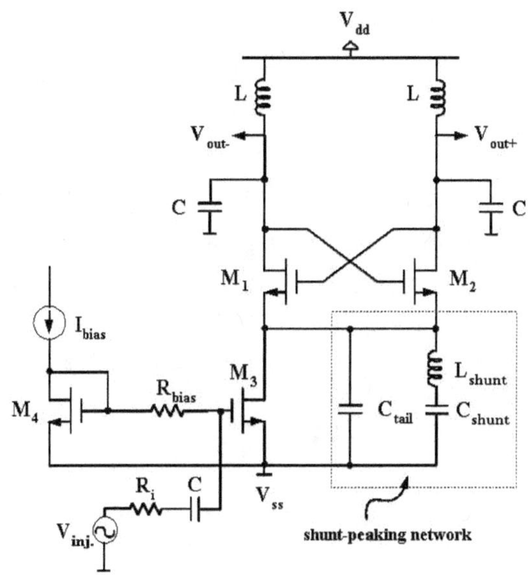

Fig. 45. Shunt-peaking in a differential injection-locked frequency divider for locking range enhancement.

ternal node. Another way to interpret is to note that the tail transistor M_3 together with C_{tail} and L_{shunt} form a tuned amplifier with output power peaking at the injection frequency, thus the name shunt-peaking. The capacitor C_{shunt} in series with the inductor L_{shunt} is connected for DC blocking. Two ILFDs with input frequencies at 9GHz and 19GHz with $0.35\mu m$ CMOS technology were designed and fabricated to prove the locking-range enhancement concept. The dividers measure 0.6 X 0.5 mm^2 including the pads. The 9GHz ILFD provides a locking range of 1490MHz (17%) with the injection power of +2dBm and tail current of 1.1mA with shunt-peaking, compared to 840MHz, 2.4dBm and 1.5mA without shunt peaking. The improvement with shunt-peaking is 77%. The measurements of the 19GHz ILFD show a locking range of 1350MHz with the injection power of +5dBm and tail current of 1mA. The supply voltage is 1.2V. The phase noise measurements of the 19GHz ILFD reported in [71] clearly shows that the locked output is not affected by the poor phase noise of the free-running oscillator and follows the phase noise of the injected signal with a 6dB offset as it is at half the frequency. The reason for this is explained in the paper to be because any phase error in the ILFD oscillation is re-adjusted by the injected signal twice every cycle once the output is locked, therefore, it does not contribute to the phase noise at the divider output.

CHAPTER IV

FULLY-INTEGRATED LC Q ENHANCEMENT RF BANDPASS FILTERS IN CMOS

The wireless communication systems have become a vital part of our daily lives. The ever increasing demand towards lower cost makes the task of the wireless communication system and circuit designers more and more challenging. This demand translates into the circuit specifications as the need of designing the circuits with lower power consumption, with smaller die area but without any compromise from the final goal of higher performance, e.g. lower noise for higher signal-to-noise ratio, higher dynamic range, etc.. The existence of large interferers, spurious tones, unwanted image and carrier frequencies as well as their harmonics in the wireless communication environment mandates the use of bandpass filters with high selectivity in the front-ends of the transceivers, [7],[83]. In the state-of-the-art solutions adopted in the industry, bulky and expensive high quality off-chip passive filters, SAW filters are used due to the demanding dynamic range specifications of the standards at the RF spectrum. There is a strong motivation behind implementing the filters in fully-integrated form using relatively cheap mainstream CMOS technologies mainly from the low cost and small form factor points of view. An additional factor that may favor integrated filters is that the off-chip filters require impedance matching circuits to interface with the integrated building blocks, complicating the system design with respect to an integrated solution. The main difficulties involved in integrating the filters at GHz range is the stringent standard specifications for the selectivity and the dynamic range. Notch filters designed with Bipolar and/or BiCMOS technologies operating at GHz range were published in the literature, [84], [85] and [86]. On the other hand, although progress has been reported using Q-enhanced LC bandpass filters integrated on Silicon, [46],

[87]-[103], the feasibility of structures designed with mainstream CMOS technologies for low voltage, low power programmable fully integrated bandpass filters at frequencies higher than 2GHz is still being investigated. This issue is here addressed through a second-order bandpass filter designed and fabricated with a $0.35\mu m$ standard CMOS technology.

In Section A, the concept of Q-enhancement LC filters is revisited. The impact of the starting quality factor of the tank on the dynamic range obtainable in the filter is briefly discussed in the same section. The design considerations of a CMOS second-order bandpass filter at 2.1GHz is the content of Section C. The dynamic range issues are emphasized, nonlinearity and noise analyses are presented in detail. Two methods are used for the nonlinearity analyses: i) using power series expansion of large-signal transfer characteristics of circuit blocks with some simplifications and ii) direct calculation of nonlinear responses proposed in [104]. The results obtained from these methods are compared to the simulated results. The design trade-offs are investigated with simplified analytical expressions and simulations. The measurement results of a prototype RF filter are provided in Chapter V.

A. Q-enhancement LC filters

The basic concept in the Q-enhancement LC filters is to use a lossy LC tank for a second-order bandpass filter and to boost the Q of the filter by incorporating a negative-conductance generator for partial compensation of the loss in the tank, [105], [83]. This is conceptually shown in Fig.46(a).

Note that assuming the loss conductance G_{loss} to be dominated by that of the inductor, with the addition of the negative conductance $-G$, the quality factor of the second order circuit becomes

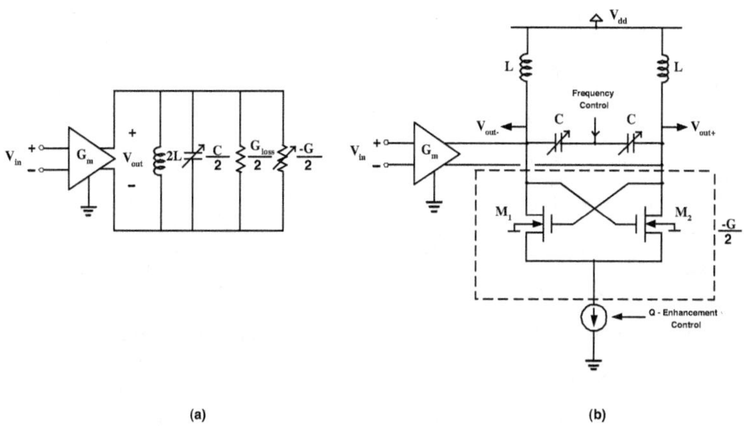

Fig. 46. Q-enhancement LC bandpass biquad: (a) concept, (b) circuit implementation in CMOS.

$$Q = \frac{Q_o}{1 - \frac{G}{G_{loss}}} = \frac{1}{\omega_o L \left(G_{loss} - G\right)} \tag{4.1}$$

where Q_o is the starting quality factor of the LC tank at resonance given by $Q_o \equiv 1/(\omega_o L G_{loss})$, [43]. Hence, the quality factor of the resonator can be enhanced to an arbitrarily high value by adjusting the negative conductance $-G$, as long as the condition $G < G_{loss}$ is observed for stability. A sample circuit implementation depicted in Fig.46(b) in a simplified form is based on the use of an LC resonator with a negative-conductance generating cross-coupled NMOS pair as the load of an input transconductance stage. The frequency tuning is through the varactors (C), whereas the Q is tuned by changing the tail current of the negative-conductance circuit.

In a Q-enhanced bandpass biquad, the 1dB compression dynamic range achievable can roughly be expressed as, [45]

$$DR = \frac{P_{1dB}}{4kT(F+1)BQ^2} \cdot Q_o^2 \qquad (4.2)$$

where P_{1dB} is the 1dB compression point of the resonator (due to the nonlinear negative conductance) in the biquad, Q_o and Q are the resonator starting quality factor (almost entirely dominated by the losses in the spiral inductor in the presented design) and the filter quality factor after the Q-enhancement, respectively, F is a noise factor associated with the active devices assuming a value on the order of 1-2, B is the -3dB bandwidth and kT is Boltzmann's constant times temperature in Kelvin degrees. The input transconductor and the frequency tuning element (e.g. varactor) in the LC tank are assumed to be linear in the calculation of the approximate expression in (4.2), leaving the negative conductance generator as the only source of nonlinearity. A similar expression is provided in [43]. The strong dependance of the dynamic range on the quality factor of the inductor Q_o is clearly seen in (4.2). The higher the starting quality factor of the inductor, the lower the noise it generates, improving the dynamic range of the filter. Furthermore, as the quality factor of the inductor increases, a lower negative conductance $-G$ is required from the cross-coupled pair for a given filter Q dictated by the selectivity requirements of the system the filter is designed for. This in turn decreases the mean square noise current generated by the negative conductance generator, as it will be clear in the noise analysis to be presented in the next section, reducing the noise factor further in addition to the reduced noise contribution from the inductor. Another advantage of using an inductor with a higher quality factor is a reduction in the power consumption as the required negative conductance and thus the required tail current in the cross-coupled differential pair is decreased. The implications of the filter quality factor Q on the dynamic range will be investigated in the following sections.

B. Circuit Design

The concept in this Q-enhancement LC filter is to use a lossy LC tank as a second-order bandpass filter with adjustable Q by moving the poles arbitrarily in the left half-plane through introducing negative conductance, the amount of which is dependent on an off-chip bias. A schematic of the bandpass filter is shown in Fig.47.

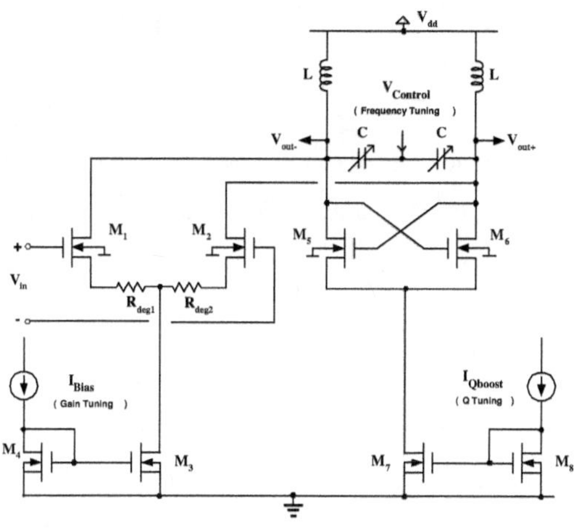

Fig. 47. Schematic of the CMOS bandpass filter.

The circuit has negative conductance added to the lossy LC tank as the load of an input transconductance stage comprised of a source-degenerated source-coupled pair. The frequency tuning is through the C (the varactors) whereas the filter Q is tuned by changing the tail current of the negative conductance circuit built with two cross-coupled NMOS transistors. Additional programmability in peak gain is incorporated into this design, which is not exploited in the previous integrated RF filters published. The peak gain programmability is controlled by the bias current of the input G_m stage. The magnitude of the voltage gain of the filter can be expressed as

$$|H(j\omega)| \cong \left| -\frac{G_{m_1}/\left(1+G_{m_1}Z_{deg}(j\omega)\right)}{G_{loss}-G+j\omega C_{tot}-j\frac{1}{\omega L}} \right| \tag{4.3}$$

where $Z_{deg}(j\omega)$ is the source degeneration impedance which is the parallel combination of R_{deg} and the equivalent capacitance at the source node of the input differential pair transistor, G_{m_1} is the transconductance of the input differential pair transistor and C_{tot} is the total output capacitance of the filter including the capacitance of the varactor denoted as C. The center frequency of the filter can be approximated as the frequency where the inductor and the total capacitance resonate, $\omega_o \cong 1/\sqrt{LC_{tot}}$. Therefore, the peak gain of the filter, i.e. the gain at the center frequency of the filter, would be

$$|H(j\omega_o)| \cong \left| -\frac{G_{m_1}/\left(1+G_{m_1}Z_{deg}(j\omega_o)\right)}{G_{loss}-G} \right|. \tag{4.4}$$

The peak gain is programmable through modifying the input transconductance G_{m_1} with the bias current of the input differential pair, as clearly seen in (4.4). It is possible to incorporate the quality factor of the filter Q into the peak gain expression in (4.4). As (4.1) suggests, with the addition of the negative conductance $-G$ for Q-enhancement, the Q of the filter is given as

$$Q = \frac{1}{\omega_o L\left(G_{loss}-G\right)}. \tag{4.5}$$

Solving (4.5) for the equivalent single-ended output conductance of the filter at the resonance, $(G_{loss} - G)$, and substituting into (4.4) yields

$$|H(j\omega_o)| \cong \left| -\frac{G_{m_1}}{\left(1+G_{m_1}Z_{deg}(j\omega_o)\right)} Q\omega_o L \right|. \tag{4.6}$$

Note in (4.6) that, increasing the quality factor of the filter through the bias of the negative conductance generator also increases the peak gain. On the other hand, once the center frequency ω_o and the Q of the filter are fixed, the peak gain can still be modified through the input transconductance G_{m_1}. Providing gain at the center frequency of an image-reject filter, for example, will be useful in a receiver front-end after the LNA, to relax the noise figure specification of the following mixer. The input G_m stage imposes trade-offs in the design of the filter between peak gain, power consumption, noise and linearity. It is possible to reduce the power consumption of the filter by reducing the peak gain, at the expense of increasing the input-referred noise but improving the linearity to keep the dynamic range almost constant. The bandpass characteristic of the filter is due to the existence of an integrated LC resonator at the load of the input G_m stage. The integrated spiral inductors of the resonator used in the CMOS filter are built with three metal layers of the technology, connected in series. The inductor model [51] used in the simulations is shown in Fig.48.

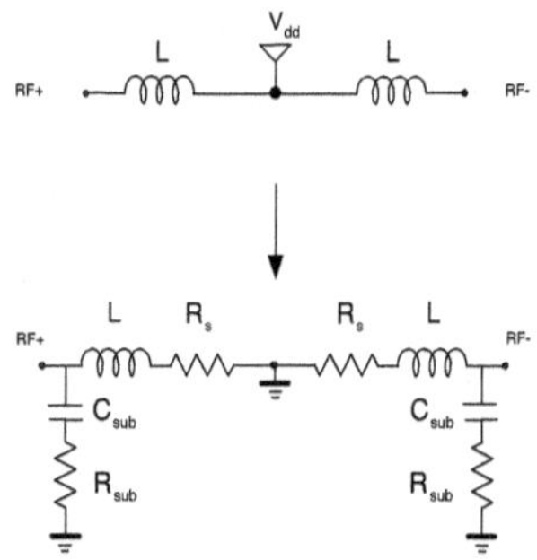

Fig. 48. Asymmetrical model of the integrated spiral inductor used in the CMOS band-pass filter simulations.

It is a simplified asymmetrical model. The cross-coupled negative resistance generator is connected to the third metal layer of each inductor to minimize the capacitive coupling to the substrate. The port where the negative resistance generator is connected and the RF signal swings (the nodes marked as $RF+$ and $RF-$ in Fig.48), is modeled with a capacitor in series with a resistor. The capacitor represents the capacitance between the third metal layer and the substrate and the resistor represents the substrate resistance. As the other port of the inductor is a common-mode node, i.e. AC ground, in the circuit, no such modeling is done at that port, for the sake of simplicity. AC simulation of the inductor model predicts a quality factor of around 2 at 2.1GHz for the inductors of 3.7nH each.

The varactors which are used to tune the center frequency of the filter are PMOS capacitors operating in depletion and inversion regions within the tuning range. Both the drain and the source terminals of the PMOS transistor are connected to its well and the capacitance looking from the gate terminal is tuned from this node as depicted in Fig.49.

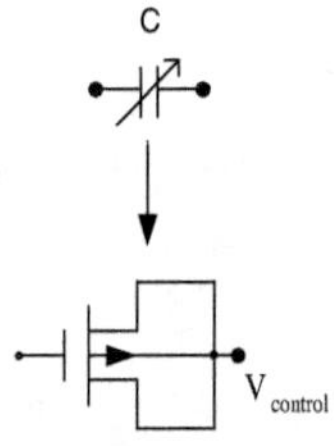

Fig. 49. PMOS capacitor in inversion-mode as the frequency tuning element.

The capacitance of the structure is tuned by modifying the inversion level in the channel with the majority carriers in the source and drain regions through changing the control voltage. The quality factor of the PMOS capacitor changes between 20, 40 within the tuning range in the AC simulations where both the well and the gate resistances are explicitly modeled, as shown in Fig.50. Note that the x-axis is the control voltage applied to the node where the drain, source and the bulk of the transistor are connected, while the gate voltage is kept constant at the dc operating point of the output of the RF filter. The capacitance decreases as the channel approaches the deep depletion, making a minimum and then the inversion takes place gradually increasing the capacitance finally reaching the value of the gate oxide capacitance. Representing the loss of the capacitor with a resistor R_C in series, the quality factor of the capacitor varies in accordance with $1/(\omega C R_C)$, reaching a maximum around the bias corresponding to the minimum capacitance in the deep depletion, as shown in Fig.50.

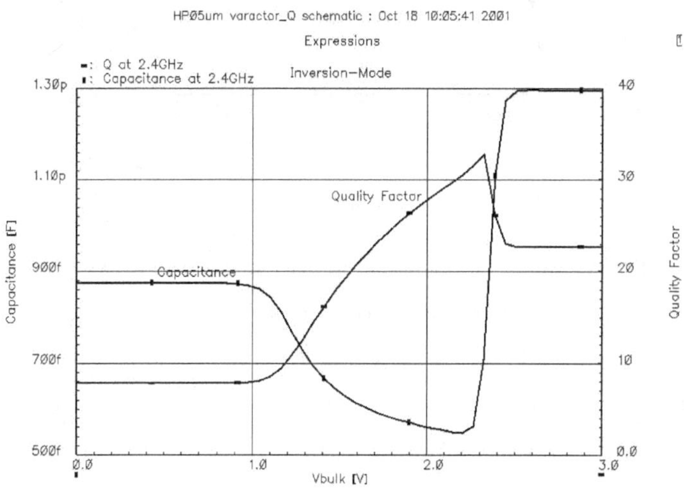

Fig. 50. Simulated capacitance-voltage transfer characteristic and the quality factor of the PMOS capacitor in inversion-mode.

The noise contributions of the components in the filter are simulated to be approximately 41%, 37%, 13%, 3.6% and 2.4%, from the lossy resonator inductors, NMOS cross-coupled transistors, input G_m transistors, source degeneration resistors and the impedance matching components, respectively. The strong dependance of the achievable dynamic range on the inductor quality factor as predicted by 4.2 is shown in Fig.51.

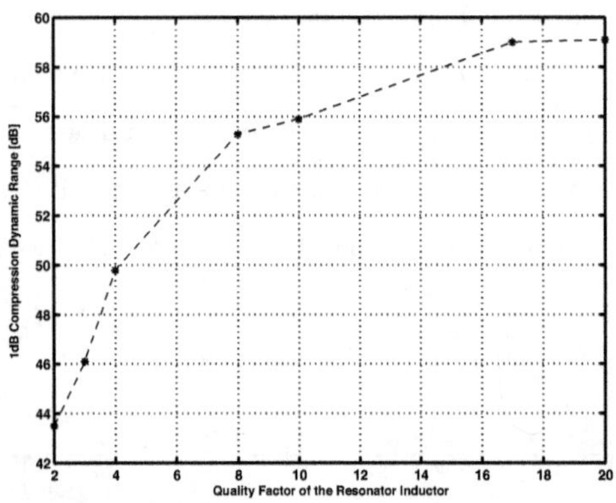

Fig. 51. Simulated 1dB compression point dynamic range versus the quality factor of the inductor with the filter Q=40 and f_o=2.28GHz.

The plot was obtained through SpectreRF simulations of the test setup including the filter, output buffer, package model, baluns and impedance matching circuitry. The center frequency and the filter Q were kept constant at f_o=2.28GHz and Q=40, respectively, in the simulations. The results reveal that by using a spiral inductor with a quality factor of 20, the dynamic range may be improved by around 15dB. The approximate expression in (4.2) predicts the improvement as 20dB. In addition, a power consumption reduction of around 50% is obtained due to the lower negative conductance required to compensate for the loss of the tank with a higher quality factor to obtain the fixed filter Q. It is also possible to improve the dynamic range by linearizing the negative conductance circuit with source-degeneration at the expense of more power consumption, as will be elaborated in the nonlinearity analysis that follows.

1. Nonlinearity Considerations

This subsection is dedicated to the study of nonlinearity in the Q-enhanced filter. The aim is to obtain easily interpretable expressions yielding the trade-offs involved in the nonlinearity performance of the design. Although a rigorous analysis making use of the approach explained in [104], which is a derivative of Volterra Series method, is possible, the resulting expressions taking into account the nonlinearities of all the contributors would be very complicated and difficult to obtain insight from. Therefore, the analyses that follow consider each nonlinearity contributor separately. There are three main contributors to the nonlinearity of the filter: the negative conductance generator, the input G_m stage and the varactor that provides the frequency tuning. Note that although the nonlinearities from each contributor interact with each other to determine the overall nonlinearity performance of the circuit, isolating each contributor serves the purpose of obtaining interpretable simplified expressions to clarify the design trade-offs involved.

a. Contribution of the Negative Conductance Generator

In analyzing the nonlinearity contribution of the negative conductance generator, MOS Level 1 equation for the drain current given in (4.7) is used for the sake of simplicity

$$I_D = \frac{\beta}{2} \cdot \frac{W}{L} \cdot (V_{GS} - V_T)^2, \qquad (4.7)$$

where β is the transconductance parameter, W and L are the width and the length of the gate of the transistor, V_{GS} is the gate-source voltage and V_T is the threshold voltage. Note that the equation allows only for the inclusion of the nonlinearity due to the transconductance g_m of the transistors. The schematic of the cross-coupled

NMOS differential pair that generates the differential negative conductance is shown in Fig.52.

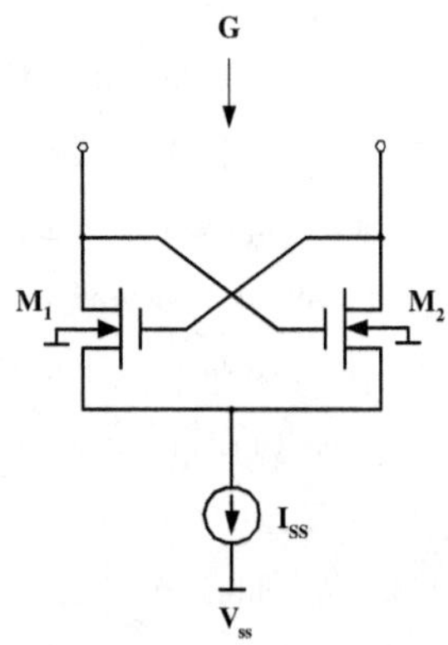

Fig. 52. Cross-coupled differential pair as negative conductance generator.

The circuit generates differential negative conductance through positive feedback. The value of the negative conductance is modified through the tail current, I_{SS}. Two equivalent small signal circuits of the cross-coupled differential pair are depicted in Fig.53. The small signal circuit shown in Fig.53(b) with its two separate voltage sources providing the input excitation is more appropriate for the large signal analysis as it will be clear in the following discussions.

From the small signal circuit in Fig.53(b), one can express the currents flowing from the voltage sources as

$$\begin{aligned} i_1 &= g_m v_2 \\ i_2 &= g_m v_1 \end{aligned} \tag{4.8}$$

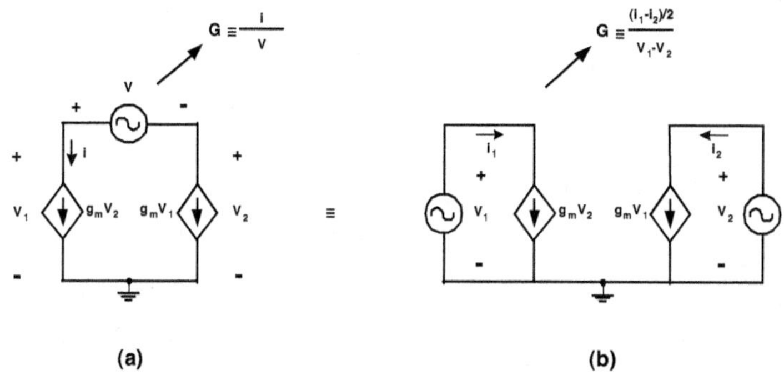

Fig. 53. Two equivalent small signal circuits of the cross-coupled differential pair.

where g_m is the transconductance of the cross-coupled transistors, which are assumed to be identical. Note that the node where the source terminals of the transistors are coupled is taken as ground because of the differential excitation of the circuit. Using the definition of the differential conductance as given in Fig.53(b) together with (4.8), the following derivation can be carried out:

$$
\begin{aligned}
G &= \frac{(i_1 - i_2)/2}{v_1 - v_2} = \frac{(g_m v_2 - g_m v_1)/2}{v_1 - v_2} \\
&= -\frac{g_m}{2} \, .
\end{aligned}
\qquad (4.9)
$$

As the derivation above shows, a cross-coupled differential pair generates a negative differential conductance with a value of half of the transconductance of each transistor. The lower the quality factor of the integrated LC tank in the filter, the higher the loss associated with it and the larger the value of the negative conductance for a given filter quality factor. One should simply increase the tail current I_{SS} of the cross-coupled differential pair to increase the value of the negative conductance.

The circuit that is used for the large signal analysis of the cross-coupled differential pair is shown in Fig.54. Note that the same input excitation and the same definition for the differential conductance as those in Fig.53(b) are used. The derivation for obtaining the large signal nonlinear differential conductance with the assumption of identical devices follows.

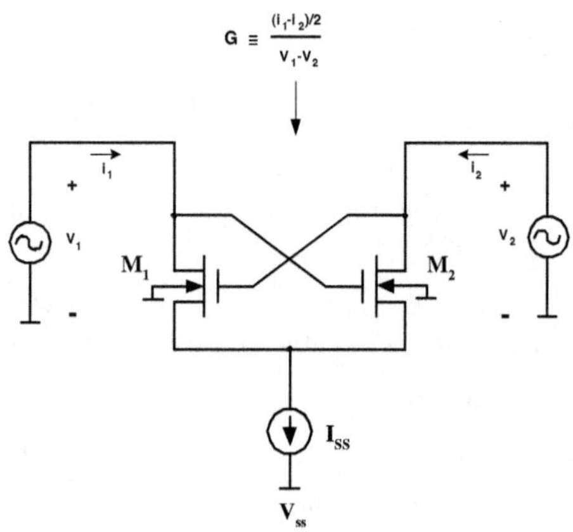

Fig. 54. Cross-coupled differential pair for analysis of the large signal negative conductance.

The node equations for the branch currents and the node voltages can be written from Fig.54 as

$$\begin{aligned} I_1 + I_2 &= I_{SS} \\ -V_1 + V_{GS_2} &- V_{GS_1} + V_2 = 0 \\ \Rightarrow V_1 - V_2 &= V_{GS_2} - V_{GS_1}. \end{aligned} \tag{4.10}$$

Substituting the gate-source voltages of the transistors based on MOS Level 1 model in (4.7), the following expression is obtained for the differential voltage across the circuit,

$$\begin{aligned} V_1 - V_2 &= \sqrt{\frac{2I_2}{\beta}} + V_T - \sqrt{\frac{2I_1}{\beta}} - V_T \\ &= \sqrt{\frac{2}{\beta}}\left(\sqrt{I_2} - \sqrt{I_1}\right). \end{aligned} \tag{4.11}$$

Let $V \equiv V_1 - V_2$, then (4.11) can be rearranged to yield

$$\sqrt{I_2} - \sqrt{I_1} = \sqrt{\frac{\beta}{2}}V. \tag{4.12}$$

Squaring each side of (4.12) produces

$$\begin{aligned} \underbrace{I_2 + I_1}_{I_{SS}} - 2\sqrt{I_1 I_2} &= \frac{\beta}{2}V^2 \\ \Rightarrow 2\sqrt{I_1 I_2} &= I_{SS} - \frac{\beta}{2}V^2 \\ I_1 I_2 &= \left(\frac{I_{SS}}{2} - \frac{\beta}{4}V^2\right)^2. \end{aligned} \tag{4.13}$$

Note that,

$$(I_2 - I_1)^2 = (I_1 + I_2)^2 - 4I_1 I_2. \tag{4.14}$$

Substituting (4.13) into (4.14) yields

$$\begin{aligned} (I_2 - I_1)^2 &= I_{SS}^2 - 4\left(\frac{I_{SS}}{2} - \frac{\beta}{4}V^2\right)^2 \\ (I_2 - I_1)^2 &= I_{SS}^2 - \left(I_{SS} - \frac{\beta}{2}V^2\right)^2. \end{aligned} \tag{4.15}$$

Taking the square root of each side in (4.15), the following is obtained:

$$I_2 - I_1 = \sqrt{I_{SS}^2 - \left(I_{SS} - \frac{\beta}{2}V^2\right)^2}. \tag{4.16}$$

Finally, factorizing the argument of the square root at the right hand side of (4.16) in accordance with $a^2 - b^2 = (a-b)(a+b)$, gives

$$I_2 - I_1 = \sqrt{\frac{\beta}{2}V^2\left(2I_{SS} - \frac{\beta}{2}V^2\right)}. \tag{4.17}$$

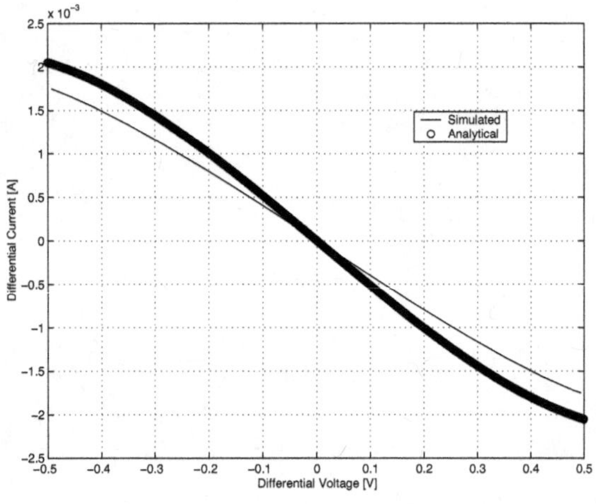

Fig. 55. Differential conductance transfer characteristic of the nonlinear negative conductance generator with $I_{SS} = 4mA$ and $V_{GS} - V_T = 385mV$.

The transconductance parameter of the transistors, β, can be expressed in terms of the quiescent drain current, $I_{SS}/2$, and the quiescent gate-source voltage, V_{GS}, of the transistors as

$$\beta = \frac{I_{SS}}{(V_{GS} - V_T)^2}. \tag{4.18}$$

Substituting (4.18) into (4.17), one obtains

$$I_2 - I_1 = \sqrt{\frac{I_{SS}V^2}{2(V_{GS}-V_T)^2}\left(2I_{SS} - \frac{I_{SS}V^2}{2(V_{GS}-V_T)^2}\right)}$$
$$I_2 - I_1 = \frac{I_{SS}(V_1 - V_2)}{(V_{GS}-V_T)}\sqrt{1 - \frac{(V_1-V_2)^2}{4(V_{GS}-V_T)^2}}\,. \quad (4.19)$$

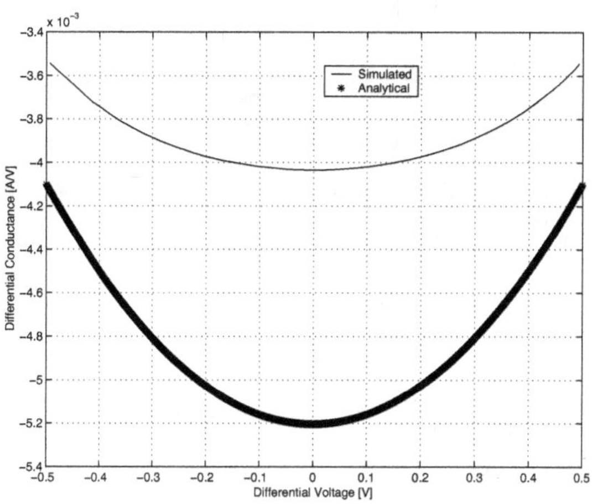

Fig. 56. Differential conductance vs. the differential voltage of the nonlinear negative conductance generator with $I_{SS} = 4mA$ and $V_{GS} - V_T = 385mV$.

The justification of using $(I_2 - I_1)^2$ instead of mathematically identical $(I_1 - I_2)^2$ at the left hand side of (4.14) and the following equations can be seen in (4.19) and by examining the circuit in Fig.54. Note that increasing $V_1 - V_2$ increases $I_2 - I_1$ as (4.19) suggests. The expression in (4.19), which models the nonlinear behavior of the negative conductance generator, is valid for $V_1 - V_2 < 2(V_{GS} - V_T)$, i.e. for weak nonlinearity. As the definition of the conductance in Fig.54 requires, (4.19) can be rearranged to obtain

$$\frac{I_1 - I_2}{2} = -\frac{1}{2}\frac{I_{SS}(V_1 - V_2)}{(V_{GS} - V_T)}\sqrt{1 - \left(\frac{V_1 - V_2}{2(V_{GS} - V_T)}\right)^2}. \tag{4.20}$$

Let $x \equiv \frac{V_1 - V_2}{2(V_{GS} - V_T)}$, then (4.20) becomes

$$\frac{I_1 - I_2}{2} = -\ I_{SS}\ x\sqrt{1 - x^2}. \tag{4.21}$$

It is possible to represent $\sqrt{1 - x^2}$ with its Taylor series expansion as, [106],

$$\sqrt{1 - x^2} = 1 - \frac{1}{2}x^2 - \frac{1}{8}x^4 \cdots \qquad -1 < x \leq 1. \tag{4.22}$$

Substituting (4.22) into (4.21), rearranging the resulting equation and using the first two terms in the Taylor Series Expansion for weak nonlinearities, the following expression is obtained for the nonlinear large signal behavior of the negative conductance generator:

$$\frac{I_1 - I_2}{2} = -\frac{I_{SS}}{2(V_{GS} - V_T)}(V_1 - V_2) + \frac{I_{SS}}{16(V_{GS} - V_T)^3}(V_1 - V_2)^3. \tag{4.23}$$

Note in (4.23) that the linear conductance term and the third-order term are

$$\begin{aligned} G &= -\frac{I_{SS}}{2(V_{GS} - V_T)} = -\frac{g_m}{2} \\ K_{3_G} &= \frac{I_{SS}}{16(V_{GS} - V_T)^3}\ . \end{aligned} \tag{4.24}$$

The comparisons between the simulated differential current and differential conductance versus differential voltage of the cross-coupled pair and the calculated characteristics with (4.23) are shown in Figs.55-56. The difference is mostly due to the simplification of using MOS Level 1 model in the derivation. The nonlinearity of the

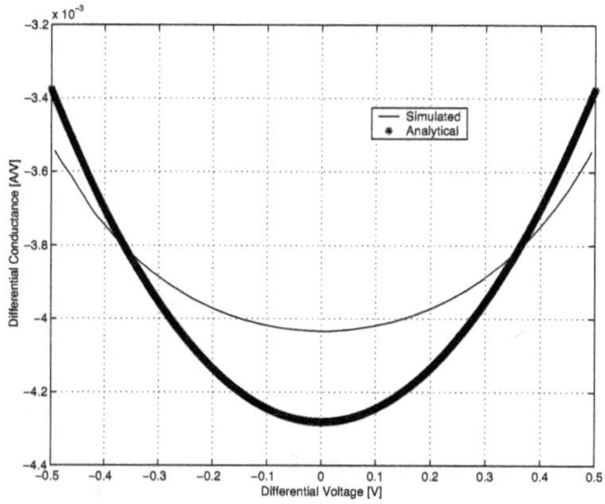

Fig. 57. Comparison of the differential current - differential voltage transfer characteristics using the simulated g_m value in lieu of the $2I_D/(V_{GS} - V_T)$ in the analytical expression, with $I_{SS} = 4mA$ and $V_{GS} - V_T = 385mV$.

output conductance of the differential pair transistors, the nonlinearity effect of the mobility reduction and the velocity saturation are ignored for the sake of obtaining simplified expressions. As it is seen in (4.24), the transconductance of the transistors is approximated with $g_m \approx I_{SS}/(V_{GS} - V_T)$, which gives 10.4mA/V as opposed to the 8.55mA/V obtained from the simulations using BSIM3 models. A better match is observed if the simulated transconductance value is used in lieu of $I_{SS}/(V_{GS} - V_T)$ in (4.23), as shown in Figs.57-58.

The incorporation of the nonlinearity of the negative conductance into the filter is conceptually illustrated in the single-ended equivalent circuit shown in Fig.59, where the negative conductance generator is represented as a nonlinear voltage-controlled current source.

The voltage-controlled current source with the value of $G_{m_{in}} v_{in}$ models the input transconductance stage and the effect of the source-degeneration resistor is embed-

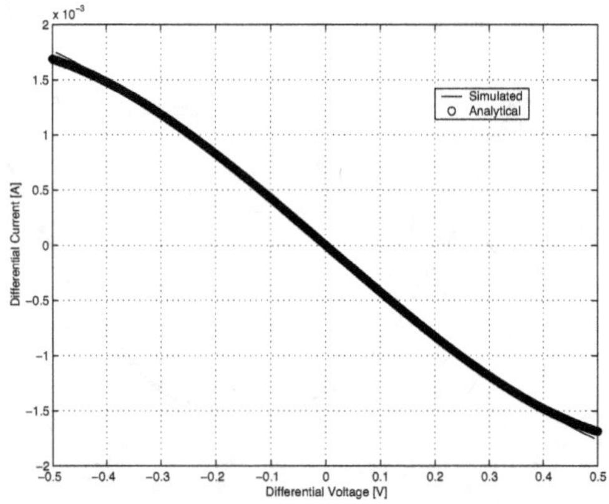

Fig. 58. Comparison of the differential conductance transfer characteristics using the simulated g_m value in lieu of the $2I_D/(V_{GS}-V_T)$ in the analytical expression with $I_{SS}=4mA$ and $V_{GS}-V_T=385mV$.

ded into it. C represents the total capacitance at the output including the varactor capacitance. G_{loss} in the equivalent circuit models the loss conductance of the tank, which is dominated by the loss from the spiral inductor. As evident in Fig.59, both the input transconductance stage and the varactor in the tank are assumed to be linear in the analysis leaving the negative conductance generator as the only source of nonlinearity at the output. The output voltage swinging across the negative conduc-

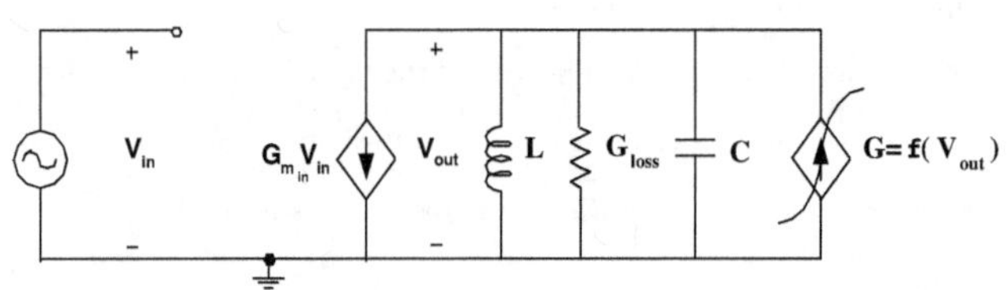

Fig. 59. Equivalent circuit of the filter for nonlinearity contribution analysis of the negative conductance generator.

tance generator produces a nonlinear current. The difference between this nonlinear current and the linear current generated by the input transconductance stage multiplied by the impedance of the lossy tank comprised of L, C and G_{loss}, generates the nonlinear voltage at the output. Obviously, there is a feedback in the circuit, i.e. the output voltage with which the negative conductance generator produces the nonlinear current is nonlinear itself, considerably complicating the analysis. The direct calculation of nonlinear responses method explained in [104] will be used to obtain an interpretable expression for input 1dB compression voltage for the case of the negative conductance generator being the only nonlinearity contributor in the filter. This analysis will be presented in a separate subsection in the following pages. In the meanwhile, based on the simplified derivation for the large signal transfer function of the negative conductance generator presented above, it should suffice to point out that the third-order nonlinearity coefficient of the negative conductance generated by the cross-coupled transistors has a strong inverse dependency to the transistors' effective bias $V_{GS} - V_T$, as expressed in (4.24). The effective bias of the cross-coupled negative conductance transistors should be increased to linearize the transfer function and to make the input 1dB compression point higher.

Adding source-degeneration resistors to the negative conductance generating cross-coupled transistors is another possibility to linearize their transconductance. This would come at the expense of extra power consumption. The higher the source degeneration factor $g_m R_{deg}$ (where g_m and R_{deg} are the transconductance of the cross-coupled transistors and the source-degeneration resistor, respectively), the better the linearity, but the higher the power consumption to obtain a given filter quality factor with a fixed integrated tank.

b. Contribution of the Input G_m Stage

The input G_m stage is an NMOS differential pair with source-degeneration resistors. The effect of the source-degeneration resistors will be taken into account after the large signal analysis of the differential pair is done. The schematic of the differential pair used in the large-signal analysis for nonlinearity study of the filter is shown in Fig.60.

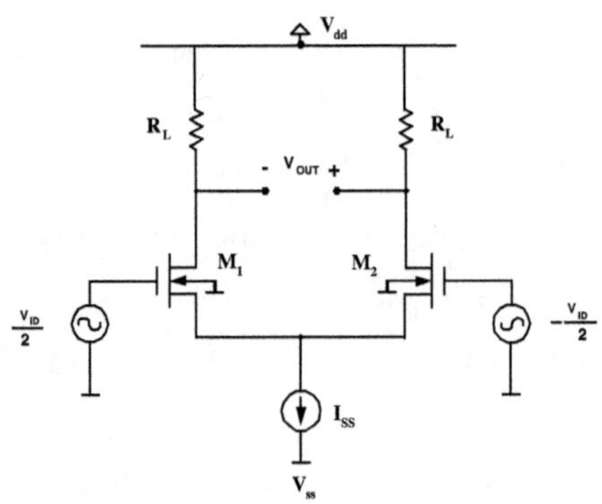

Fig. 60. Differential pair for the analysis of the large signal transconductance.

The derivation for obtaining the large signal nonlinear differential transconductance with the assumption of identical devices follows. It is very similar to the derivation presented before for the cross-coupled differential pair. The node equations for the branch currents and the node voltages can be written from Fig.60 as

$$\begin{aligned} I_{D_1} + I_{D_2} &= I_{SS} \\ -\frac{V_{ID}}{2} + V_{GS_1} &- V_{GS_2} - \frac{V_{ID}}{2} = 0 \\ \Rightarrow V_{GS_1} - V_{GS_2} &= V_{ID}. \end{aligned} \tag{4.25}$$

Substituting the gate-source voltages of the transistors based on MOS Level 1 model in (4.7), the following expression is obtained for the differential input voltage of the circuit,

$$\begin{aligned} V_{ID} &= \sqrt{\frac{2I_{D_1}}{\beta}} + V_T - \sqrt{\frac{2I_{D_2}}{\beta}} - V_T \\ &= \sqrt{\frac{2}{\beta}}\left(\sqrt{I_{D_1}} - \sqrt{I_{D_2}}\right). \end{aligned} \tag{4.26}$$

On the other hand, the output voltage can be expressed as

$$V_{OUT} = (V_{DD} - I_{D_2}R_L) - (V_{DD} - I_{D_1}R_L) = R_L(I_{D_1} - I_{D_2}). \tag{4.27}$$

Rearranging (4.26) for the difference of the square root of the drain currents and squaring each side of resulting equation produces

$$\begin{aligned} \underbrace{I_{D_1} + I_{D_2}}_{I_{SS}} - 2\sqrt{I_{D_1}I_{D_2}} &= \frac{\beta}{2}V_{ID}^2 \\ \Rightarrow 2\sqrt{I_{D_1}I_{D_2}} &= I_{SS} - \frac{\beta}{2}V_{ID}^2 \\ I_{D_1}I_{D_2} &= \left(\frac{I_{SS}}{2} - \frac{\beta}{4}V_{ID}^2\right)^2. \end{aligned} \tag{4.28}$$

Note that,

$$(I_{D_1} - I_{D_2})^2 = (I_{D_1} + I_{D_2})^2 - 4I_{D_1}I_{D_2}. \tag{4.29}$$

Substituting (4.28) into (4.29) yields

$$(I_{D_1} - I_{D_2})^2 = I_{SS}^2 - 4\left(\frac{I_{SS}}{2} - \frac{\beta}{4}V_{ID}^2\right)^2$$

$$(I_{D_1} - I_{D_2})^2 = I_{SS}^2 - \left(I_{SS} - \frac{\beta}{2}V_{ID}^2\right)^2. \tag{4.30}$$

Taking the square root of each side in (4.30), the following is obtained:

$$I_{D_1} - I_{D_2} = \sqrt{I_{SS}^2 - \left(I_{SS} - \frac{\beta}{2}V_{ID}^2\right)^2}. \tag{4.31}$$

Factorizing the argument of the square root at the right hand side of (4.31) in accordance with $a^2 - b^2 = (a-b)(a+b)$, gives

$$I_{D_1} - I_{D_2} = \sqrt{\frac{\beta}{2}V_{ID}^2\left(2I_{SS} - \frac{\beta}{2}V_{ID}^2\right)}. \tag{4.32}$$

Substituting (4.32) into (4.27) yields

$$V_{OUT} = R_L\sqrt{\frac{\beta}{2}V_{ID}^2\left(2I_{SS} - \frac{\beta}{2}V_{ID}^2\right)}. \tag{4.33}$$

The transconductance parameter of the transistors, β, can be expressed in terms of the quiescent drain current, $I_{SS}/2$, and the quiescent gate-source voltage, V_{GS}, of the transistors as

$$\beta = \frac{I_{SS}}{(V_{GS} - V_T)^2}. \tag{4.34}$$

Substituting (4.34) into (4.33), one obtains

$$\begin{aligned} V_{OUT} &= R_L\sqrt{\frac{I_{SS}V_{ID}^2}{2(V_{GS}-V_T)^2}\left(2I_{SS} - \frac{I_{SS}V_{ID}^2}{2(V_{GS}-V_T)^2}\right)} \\ V_{OUT} &= \frac{R_L V_{ID} I_{SS}}{(V_{GS}-V_T)}\sqrt{1 - \frac{V_{ID}^2}{4(V_{GS}-V_T)^2}}. \end{aligned} \tag{4.35}$$

The expression in (4.35), which models the nonlinear behavior of the differential pair, is valid for $|V_{ID}| < 2(V_{GS} - V_T)$, i.e. for weak nonlinearity. Note that the first term at the right-hand side is the linear term valid for small-signal excitation:

$$v_{OUT} = g_m R_L v_{ID} = \underbrace{\frac{2\frac{I_{SS}}{2}}{(V_{GS} - V_T)}}_{2I_D/(V_{GS}-V_T)} R_L v_{ID}. \tag{4.36}$$

Let $x \equiv \frac{V_{ID}}{2(V_{GS}-V_T)}$, then (4.35) becomes

$$V_{OUT} = 2R_L I_{SS} x\sqrt{1-x^2}. \tag{4.37}$$

It is possible to represent $\sqrt{1-x^2}$ with its Taylor series expansion as, [106],

$$\sqrt{1-x^2} = 1 - \frac{1}{2}x^2 - \frac{1}{8}x^4 \cdots \quad -1 < x \leq 1. \tag{4.38}$$

Substituting (4.38) into (4.37), rearranging the resulting equation and using the first two terms in the Taylor Series Expansion for weak nonlinearities, the following expression is obtained for the nonlinear large signal behavior of the differential pair:

$$V_{OUT} \approx R_L I_{SS} \left(\frac{V_{ID}}{(V_{GS} - V_T)} - \frac{V_{ID}^3}{8(V_{GS} - V_T)^3} \right). \tag{4.39}$$

Note in (4.39) that the linear and the third-order terms are

$$\begin{aligned} K_1 &= \frac{R_L I_{SS}}{(V_{GS} - V_T)} \\ K_3 &= -\frac{R_L I_{SS}}{8(V_{GS} - V_T)^3}. \end{aligned} \tag{4.40}$$

Third-order harmonic distortion is defined as, [104],

$$HD_3 \equiv \frac{1}{4}A^2 \left| \frac{K_3}{K_1} \right| \tag{4.41}$$

where A denotes the amplitude of the input signal. Substituting (4.40) into (4.41), together with V_{ID} in lieu of A, yields

$$HD_3 \equiv \frac{1}{4}V_{ID}^2 \left| \frac{V_{GS} - V_T}{8(V_{GS} - V_T)^3} \right| = \frac{1}{32} \frac{V_{ID}^2}{(V_{GS} - V_T)^2} \cdot \tag{4.42}$$

For small input amplitudes and at low frequencies, third-order intermodulation distortion, IM_3, is three times the third-order harmonic distortion, [104].

$$\Rightarrow IM_3 = \frac{3}{32} \frac{V_{ID}^2}{(V_{GS} - V_T)^2} \cdot \tag{4.43}$$

In order to calculate the third-order intercept point, IP_3, which is defined to be the input amplitude that generates a third-order intermodulation product equal to the fundamental component, IM_3 should be equalized to 1 and the corresponding input amplitude V_{ID} should be solved in (4.43) to produce

$$IP_3 = 4\sqrt{\frac{2}{3}}(V_{GS} - V_T) \cdot \tag{4.44}$$

Note in (4.42)-(4.44) that all these equations quantifying the nonlinearity in a differential pair show that increasing the effective bias of the transistors improves the nonlinearity performance. Recall that similar dependency to the effective bias of the transistors was pointed out in the previous section where the nonlinearity contribution of the negative conductance generator was studied.

The differential pair used as the input G_m stage of the filter has source degeneration resistors for linearization. Using the same derivation procedure as the one above does not yield a closed-form solution in case the source-degeneration resistors are included. The effect of the resistors can be roughly approximated by using the results in Section 4.8.3.2 regarding emitter degeneration in a single-transistor amplifier in [104]. It is shown in [104] that the fundamental, second and third-order terms of the output current of the amplifier are affected by the emitter degeneration resistor as follows:

$$
\begin{aligned}
I_{OUT} &= g_m V_{in} + K_{2_{gm}} V_{in}^2 + K_{3_{gm}} V_{in}^3 \quad \text{without } R_E \\
I_{OUT}^{'} &= g_m^{'} V_{in} + K_{2_{gm}}^{'} V_{in}^2 + K_{3_{gm}}^{'} V_{in}^3 \quad \text{with } R_E
\end{aligned} \tag{4.45}
$$

where the fundamental and the nonlinearity terms with the emitter degeneration are

$$
\begin{aligned}
g_m^{'} &= \frac{g_m}{1+g_m R_E} \\
K_{2_{gm}}^{'} &= \frac{K_{2_{gm}}}{(1+g_m R_E)^3} \\
K_{3_{gm}}^{'} &= \left(K_{3_{gm}} - 2K_{2_{gm}}^2 R_E \frac{1}{1+g_m R_E}\right) \frac{1}{(1+g_m R_E)^4}\,.
\end{aligned} \tag{4.46}
$$

Note that, the input G_m stage is a differential circuit where the even-order harmonics are suppressed ideally (assuming no mismatches between the devices), and thus no second order term is found in the derivation presented above for the differential pair, as seen in (4.40). Adopting the results for the emitter degeneration given in (4.46) to the differential pair with source degeneration taking into account the fact that the second order term K_2 does not exist for the latter, the terms in (4.40) become

$$
\begin{aligned}
K_1^{'} &= \frac{R_L I_{SS}}{(V_{GS}-V_T)} \left(\frac{1}{1+g_m R_{deg}}\right) \\
K_3^{'} &= -\frac{R_L I_{SS}}{8(V_{GS}-V_T)^3} \left(\frac{1}{1+g_m R_{deg}}\right)^4\,.
\end{aligned} \tag{4.47}
$$

All the derivations so far are based on the differential pair with resistive load R_L as shown in Fig.60. Recall that in the filter the load of the input G_m stage is an LC tank with a negative conductance for Q boosting. The load resistance R_L in the equations simply serve to convert the output current into output voltage. Therefore, the terms in (4.47) with R_L should be modified accordingly to express the nonlinearity in the filter contributed by the input G_m stage. R_L should be substituted by the magnitude of the load impedance of the filter at either the resonance frequency, $|Z_L(j\omega_o)|$, or the third harmonic frequency, $|Z_L(j3\omega_o)|$, depending on the frequency of the current component to be converted into voltage. The magnitude of the load impedance of the filter at the resonance frequency is given as

$$|Z_L(j\omega_o)| = \left| \frac{1}{G_{loss} + G_{neg} + j\left(\omega_o C - \frac{1}{\omega_o L}\right)} \right| \approx \frac{1}{G_{loss} + G_{neg}} . \tag{4.48}$$

where G_{neg} is the single-ended negative conductance generated by the cross-coupled differential pair used for Q-boosting.

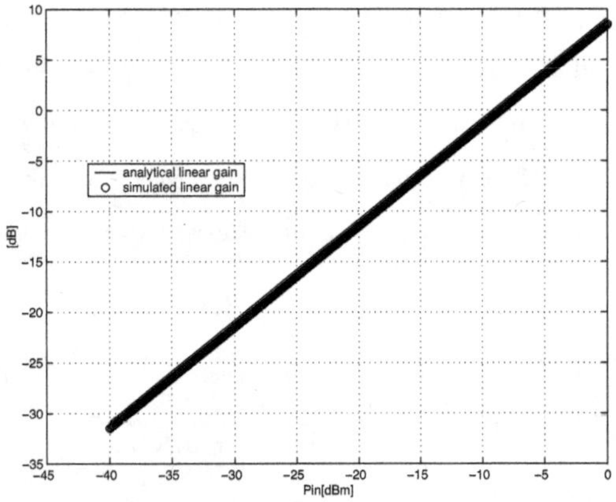

Fig. 61. Linear gain characteristic of the filter with only the nonlinearity of the input G_m stage included at $f_{in} = 2.3GHz$ and $Q = 40$.

The nonlinear output current generated by the input G_m stage is converted into a nonlinear voltage through the Q-boosted LC tank. Remember that both the negative conductance generator which boosts the Q of the tank and the varactor providing the frequency tuning are assumed to be linear in this analysis. The nonlinear output voltage generated as a response to an input signal at the resonance frequency is given by

$$V_{out} = -\underbrace{\left(G'_{m_{in}} + K'_{3G_{m_{in}}} V_{in}^2\right)}_{nonlinear\ input\ transconductance} V_{in} \left|Z_L(j\omega_o)\right| \tag{4.49}$$

where $G'_{m_{in}}$ and $K'_{3G_{m_{in}}}$ are the linear and third-order terms of the input transconductance, respectively, given as

$$\begin{aligned} G'_{m_{in}} &= \overbrace{\frac{I_{SS}}{(V_{GS}-V_T)}}^{G_{m_{in}}} \left(\frac{1}{1+G_{m_{in}}R_{deg}}\right) \\ K'_{3G_{m_{in}}} &= -\frac{I_{SS}}{8(V_{GS}-V_T)^3} \left(\frac{1}{1+G_{m_{in}}R_{deg}}\right)^4 . \end{aligned} \tag{4.50}$$

Let $V_{in} = V\cos(\omega_o t)$ for a sinusoidal input. Using the trigonometric relationship of $\cos^3(\omega_o t) = \frac{1}{4}(3\cos(\omega_o t) + \cos(3\omega_o t))$, substituting (4.48) into (4.49) and decomposing the latter into the fundamental and third harmonic terms, the following expression is obtained

$$\begin{aligned} V_{out} &= \underbrace{-\left(\overbrace{\left(\frac{G'_{m_{in}}}{G_{loss}+G_{neg}}\right)}^{linear\ term} + \overbrace{\frac{3}{4}\frac{K'_{3G_{m_{in}}} V^2}{(G_{loss}+G_{neg})}}^{nonlinear\ term}\right) V\cos(\omega_o t)}_{fundamental} \\ &\quad - \underbrace{\frac{1}{4}K'_{3G_{m_{in}}} \left|Z_L(j3\omega_o)\right| V^3 \cos(3\omega_o t)}_{third\ harmonic} . \end{aligned} \tag{4.51}$$

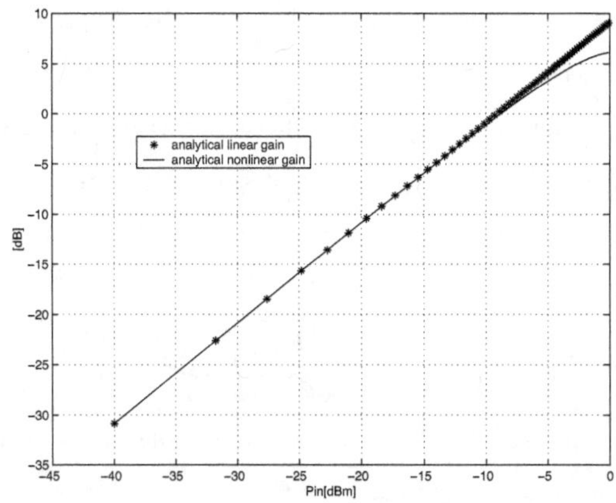

Fig. 62. Gain compression of the filter with only the nonlinearity of the input G_m stage included at $f_{in} = 2.3GHz$ and $Q = 40$.

It is vital to note in (4.51) that the output current component at the third harmonic is multiplied with the magnitude of the equivalent load impedance of the filter at the frequency of the third harmonic, $|Z_L(j3\omega_o)|$, to generate the third harmonic component at the output voltage. Once again, this allows to account for the bandpass characteristic of the transfer function that attenuates the output component at the third harmonic. A comparison between the analytical linear gain calculated using the linear term in (4.51) and the simulated gain at the resonance is shown in Fig.61. It should be repeated that the quality factor, Q, of the filter is adjusted through the value of the negative conductance to be 40 at the center frequency of around f_o = 2.3GHz both in the simulations, where only the differential pair with source degeneration is taken at transistor-level, and in the analytical calculations.

Note that the nonlinear term in the fundamental component of the output voltage in (4.51) causes the compression in the power gain as the input amplitude V increases, since $K_{3_{G_{m_{in}}}}'$ is negative, as shown in Fig.62.

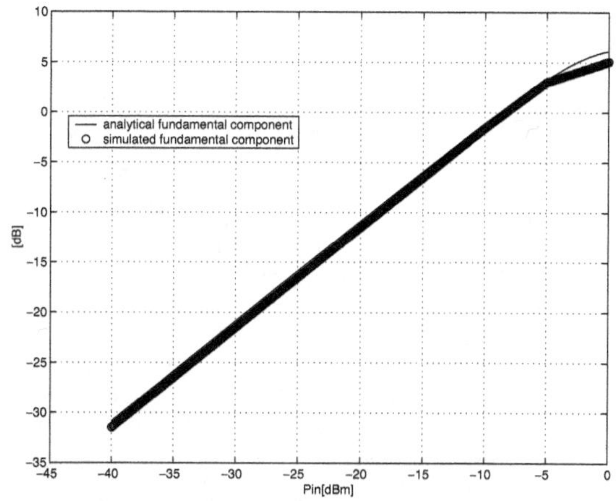

Fig. 63. Fundamental output power versus the input power of the filter with only the nonlinearity of the input G_m stage included at $f_{in} = 2.3GHz$ and $Q = 40$.

Simulated and analytically calculated fundamental component versus the input power characteristics are plotted in Fig.63. The simulations were run with SpectreRF Swept Periodic Steady State Analysis. The analytical model given in (4.51) matches with the simulation results fairly well for weak nonlinearities and it starts to diverge as the nonlinearities get stronger, as expected. On the other hand, the simulated and the calculated output power at the third harmonic versus the input power characteristics are plotted in Fig.64.

An approximate expression for the input 1dB compression point can be obtained from (4.51). Based on the definition, the following equation can be written

$$20\log_{10}\left(B\,V_{1dB}\right) - 20\log_{10}\left(\left|B + \frac{3}{4}\frac{K_{3G_{m_{in}}}^{'}\,V_{1dB}^{2}}{\left(G_{loss}+G_{neg}\right)}\right| V_{1dB}\right) = 1 \qquad (4.52)$$

where

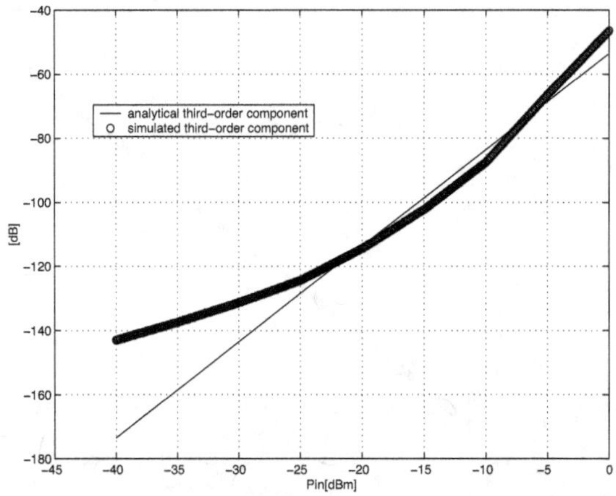

Fig. 64. Third harmonic at the output of the filter with only the nonlinearity of the input G_m stage included at $f_{in} = 2.3GHz$ and $Q = 40$.

$$B \equiv \frac{G'_{m_{in}}}{G_{loss} + G_{neg}} \tag{4.53}$$

is the magnitude of the linear gain of the filter. Solving the 1dB compression point input voltage, V_{1dB}, in (4.52) gives

$$V_{1dB} \approx 1.08 \times (V_{GS} - V_T) \times \sqrt{(1 + G_{m_{in}} R_{deg})^3} \, . \tag{4.54}$$

As (4.54) shows, the effective bias of the transistors in the differential pair should be maximized to reduce the nonlinearity contribution of the input G_m stage. Having a limit on the tail current I_{SS} due to the power budget, increasing the effective bias reduces the transconductance of the transistors with fixed dimensions, degrading the noise performance of the filter. Fixing the I_{SS} at an upper limit for power consumption concern, one way to increase the effective bias for a certain transconductance value, in order not to degrade the noise, is to reduce the W/L of the transistors, as $g_m = \beta(V_{GS} - V_T)$ suggests. On the other hand, it is possible to decrease the contribution by using source degeneration with extra power consumption for the same input transconductance.

Table V tabulates the simulated input 1dB compression points and the calculated 1dB compression points with only the nonlinearity from the input G_m stage taken into account for different filter quality factors. Comparing the values with those obtained for the case where only the nonlinearity from the negative conductance generator is taken into account (shown in Fig.68), it is clear that the contribution from the input G_m stage is much less than that of the negative conductance generator. The conclusion is also intuitive in the sense that although both the negative conductance generator and the input differential pair have comparable third order nonlinearity coefficients in magnitude, the voltage across the negative conductance generator is the output voltage of the input G_m stage and it increases as the quality factor of the filter increases. This effect makes the nonlinearity contribution of the negative conductance generator much more pronounced than that of the input G_m stage.

It should be noted that the derivations presented make the assumption of weak nonlinearity, and only the nonlinearity at the transconductance of the differential pair transistors are taken into account. The parasitic nonlinear junction capacitors, the nonlinearity of the output conductance of the differential pair transistors, the nonlinearity due to bulk-effect in the transistors and the nonlinearity effect of the mobility reduction and the velocity saturation are ignored for the sake of obtaining simplified expressions. The analytical 1dB compression point expression in (4.54) does not have any dependence to the quality factor Q of the filter, whereas the simulated values in Table V show that the 1dB compression point decreases as the filter quality factor increases. This is anticipated to be due to the nonlinearity contributions from the junction capacitors, the output conductance of the differential pair transistors and the velocity saturation, all of which would be expected to have an increasing effect as the voltage swing at the drain terminals of the transistors increase with the increased filter quality factor.

Table V. Simulated vs. calculated Input 1 dB compression point for various filter quality factors with the input G_m stage as the only nonlinearity contributor.

Filter Q	Simulated $P_{in_{1dB}}$	Calculated $P_{in_{1dB}}$
53	-8.4dBm	-4.3dBm
46	-5.2dBm	-4.3dBm
41.5	-4.1dBm	-4.3dBm
35	-2.94dBm	-4.3dBm
31.5	-2.04dBm	-4.3dBm
25	-0.13dBm	-4.3dBm
22	0.043dBm	-4.3dBm

c. Contribution of the Varactor

The frequency tuning in the filter is provided by the PMOS capacitor operated in depletion and inversion regions. The capacitance-voltage transfer characteristic of the PMOS capacitor (varactor) is highly nonlinear, contributing to the nonlinearity of the filter. This section is devoted to the investigation of the contribution of the varactor to the nonlinearity of the filter. The main aim is to obtain an approximate but interpretable expression for the 1dB compression point of the filter in case only the nonlinearity of the varactor is taken into account. The expression should provide some insights into the trade-offs involved in the design of the filter for better linearity.

The first step is to expand the nonlinear capacitance-voltage transfer characteristic of the varactor to its power series. The varactor's region of operation in the simulations is depletion and therefore the capacitance-voltage transfer characteristic is approximated by that of a nonlinear junction capacitor given as

$$C_j = \frac{C_{j0}}{\left(1 + \frac{V_j}{\phi}\right)^{m_j}} \tag{4.55}$$

where C_{j0} is the zero-bias junction capacitance, ϕ is the built-in junction potential, m_j is the junction grading coefficient and V_j is the voltage across the junction. The nonlinear relationship between the capacitance of the varactor C_V and the voltage across it, denoted as v_c, is given by

$$\begin{aligned} C_V &= f\left(v_c(t)\right) = f\left(V_C + v(t)\right) \\ &= f\left(V_C\right) + \sum_{k=1}^{\infty} \frac{1}{k!} \left.\frac{\partial^k f\left(v(t)\right)}{\partial v^k}\right|_{v=V_C} \cdot v^k(t) \end{aligned} \tag{4.56}$$

where V_C is the quiescent value of the voltage across the capacitor. Defining the second and third-order nonlinearity coefficients as, [104]

$$
\begin{aligned}
K_{2_{C_V}} &\equiv \left.\frac{dC_V}{dv}\right|_{v=V_C} \\
K_{3_{C_V}} &\equiv \left.\frac{1}{2!}\frac{d^2C_V}{dv^2}\right|_{v=V_C}
\end{aligned}
\tag{4.57}
$$

the power series representation of the varactor using up to the third-order nonlinearity coefficient around the value at V_C can be expressed as

$$
\begin{aligned}
C_V &= C_V\left(V_C\right) + K_{2_{C_V}} v + K_{3_{C_V}} v^2 + \cdots \\
&= C_V\left(V_C\right) + \left.\frac{dC_V}{dv}\right|_{v=V_C} \cdot v + \left.\frac{1}{2}\frac{d^2C_V}{dv^2}\right|_{v=V_C} \cdot v^2 + \cdots
\end{aligned}
\tag{4.58}
$$

where v is the incremental value of the voltage across the capacitor with respect to its quiescent value. Taking the derivatives in (4.57) using the model in (4.55) and substituting them into (4.58) yields

$$
C_V = \frac{C_0}{\left(1+\frac{V_C}{\phi}\right)^m} - \frac{m}{\phi}\cdot\frac{C_0}{\left(1+\frac{V_C}{\phi}\right)^{m+1}}\cdot v + \frac{m(m+1)}{2\phi^2}\cdot\frac{C_0}{\left(1+\frac{V_C}{\phi}\right)^{m+2}}\cdot v^2 + \cdots \tag{4.59}
$$

(4.59) is used to approximate the capacitance-voltage transfer characteristic of the PMOS varactor operated in accumulation-depletion. Recall that m in the equation above represents the junction grading coefficient m_j in (4.55) and its value is adjusted in order to best fit the characteristic in (4.59) to the simulated PMOS varactor characteristic. A comparison between the capacitance-voltage transfer characteristic based on (4.55), the characteristic obtained from its power series expansion given in (4.59) and the simulated capacitance-voltage transfer curve of the PMOS varactor is shown in Fig.65. The power series expansion is evaluated at $V_C = 0.9V$ as dictated by the simulations. The matching between the simulated curve and the power series expansion around $V_C = 0.9V$ depicted in Fig.65 seems reasonably well for the rough approximation targeted in the derivation of the 1dB compression voltage expression tackled in the next subsection.

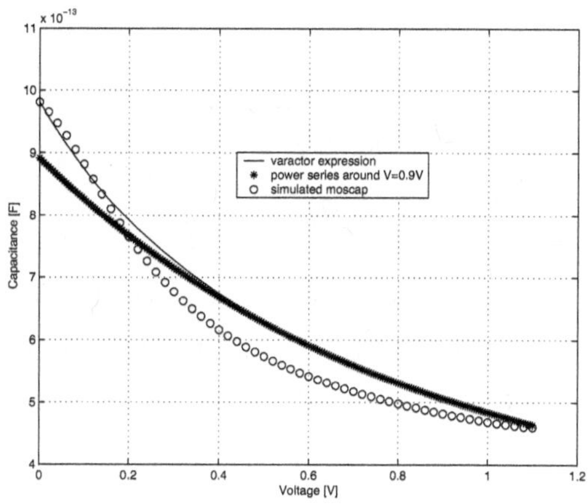

Fig. 65. Capacitance-voltage transfer characteristic of the varactor.

The incorporation of the varactor nonlinearity to the nonlinearity of the filter is based on the equivalent circuit in Fig.66.

The voltage-controlled current source with the value of $G_{m_{in}}v_{in}$ models the input transconductance stage and the effect of the source-degeneration resistor is embedded into it. C_V represents the nonlinear varactor capacitance and C is the remaining part of the total output capacitance assumed to be linear including the parasitic capacitance from the inductor and the transistors. G, given as $G_{loss}+G_{neg}$, is the equivalent conductance at the output including the loss conductance of the tank denoted as G_{loss}, which is dominated by the loss from the spiral inductor, and the negative conductance, denoted as G_{neg}. As evident in Fig.66, both the input transconductance stage

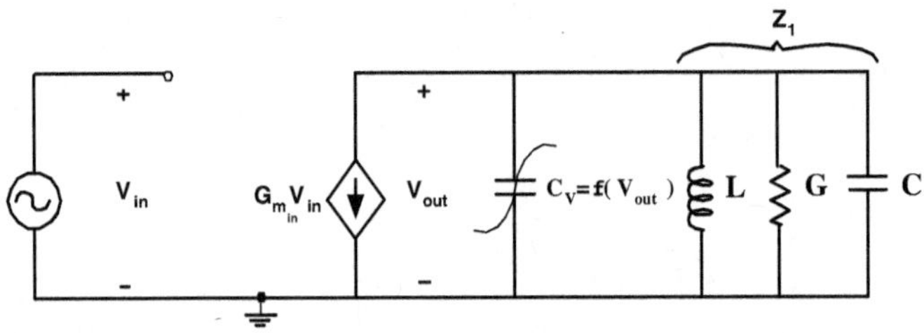

Fig. 66. Equivalent circuit of the filter for nonlinearity contribution analysis of the varactor.

and the negative conductance generator are assumed to be linear in this analysis leaving the varactor as the only source of nonlinearity at the output. The output voltage swinging across the varactor produces a nonlinear current. The difference between this nonlinear current and the linear current generated by the input transconductance stage multiplied by the impedance denoted as Z_1 in Fig.66 generates the nonlinear voltage at the output. Obviously, there is a feedback in the circuit, i.e. the output voltage with which the varactor produces the nonlinear current is nonlinear itself, considerably complicating the analysis. The direct calculation of nonlinear responses method explained in [104] will be used to obtain an interpretable expression for the input 1dB compression voltage for the case of the varactor being the only nonlinearity contributor in the filter. This analysis will be presented in the next subsection. To justify the use of the equivalent circuit given in Fig.66 as a basis of the analysis to be presented in the next subsection, one may assume the varactor to be linear and compare the magnitude of the small-signal output voltage obtained from this equivalent circuit with that from the simulations. The magnitude of the output voltage from linear small-signal analysis based on the equivalent circuit shown in Fig.66 can be expressed as

$$V_{out_{lin}} = G_{m_{in}} \left| \frac{1}{G + j\left(\omega(C + C_V) - \frac{1}{\omega L}\right)} \right| V_{in}. \tag{4.60}$$

The calculated and the simulated magnitude frequency characteristics of the filter are compared in Fig.67. The value of the negative conductance is adjusted such that the resulting quality factor of the filter is around 40 at 2.34GHz. Recall that the simulated filter is differential.

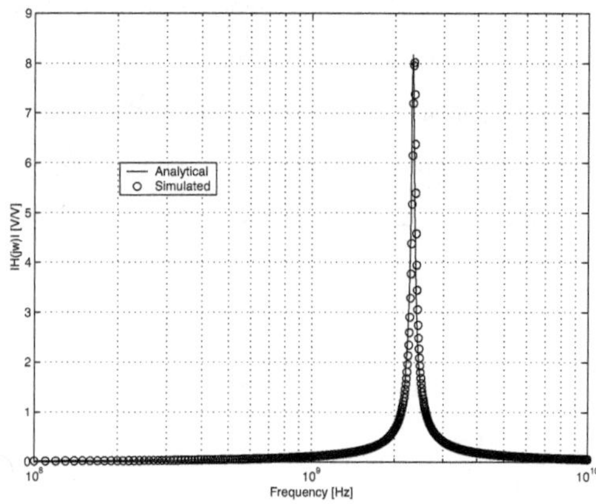

Fig. 67. Frequency characteristic of the magnitude of the small-signal voltage gain of the filter.

To conclude the qualitative discussion in this subsection, it should be noted that as the varactor is at the output of the filter, the higher the peak voltage gain at the center frequency, the larger the voltage swing across the nonlinear varactor, thus the higher its contribution to the filter nonlinearity.

d. Nonlinearity Analysis with Direct Calculation of Nonlinear Responses

In the nonlinearity analyses of the filter presented above, the discussions were based on the power series expansion of the simplified transfer characteristics of the building blocks in the circuit. Recall that in order to be able to obtain closed form expressions for the large signal characteristics of the differential pair and the negative conductance generator, MOS level 1 model was used. Some further simplifications were also made throughout the analyses. Expressions obtained from more rigorous analyses of the nonlinearity contributions of the building blocks in the circuit are presented below. The method explained in [104] is used in the derivations. It is a variant of Volterra Series and it is called as "Direct Calculation of Nonlinear Responses". The Volterra Series describe the output of a nonlinear system as the sum of a first-order operator, a second-order operator, third-order operator and so on, [104]. Every operator is described either in the time domain or in the frequency domain with a kind of transfer function called a Volterra Kernel.

The direct calculation of nonlinear responses method presented in [104] circumvents the use of Volterra series. The resulting harmonics or intermodulation products obtained with the method are identical to those obtained with Volterra Series approach, [104]. Both methods can only compute a reasonable approximation of a circuit's nonlinear behavior when the circuit behaves in a weakly nonlinear way. In practice, it is not always easy to predict the limit of the signal amplitudes up to which the weak nonlinearity assumption is valid. In order to check the validity of the assumption, iterative numerical simulation techniques must be used. A reasonable match between the simulations ran with SpectreRF and the theoretical computations with the direct calculation of nonlinear responses method validated the weak nonlinearity assumptions up to the input amplitudes which cause 1dB compression at the output. Note that expressing the 1dB compression point in terms of the circuit parameters is the main aim of the nonlinearity analyses.

The filter is a single-input circuit and the difference between the direct calculation of nonlinear responses method and the Volterra Series method for single-input circuits lies in the value of the nonlinear current sources connected in parallel with the nonlinear components in the circuit for the analysis, [104].

The expressions obtained with the direct calculation of nonlinear responses provide closer results to the simulated nonlinearities than the simplified analyses that were presented in the previous section in detail. However, it is important to note that the trade-offs that are pointed out are the same as those resulted from the simplified analyses of the previous section. Some parts in the discussion below may seem as repetitions of what were mentioned previously, but they are still included for the sake of completeness.

Although a rigorous analysis taking into account the nonlinearities of all the contributors is possible, the resulting expressions would be very complicated and difficult to obtain insight from. Therefore, the analyses presented here consider each nonlinearity contributor separately. Note that although the nonlinearities from each contributor interact with each other to determine the overall nonlinearity performance of the circuit, isolating each contributor serves the purpose of obtaining interpretable simplified expressions to yield the design trade-offs involved.

The saturation drain current equation used in the calculations accounts for the mobility reduction due to the vertical electrical field and it is given by

$$I_D = \frac{\mu_o}{1+\theta(V_{GS}-V_T)}\frac{C_{ox}}{2}\frac{W}{L}(V_{GS}-V_T)^2 \tag{4.61}$$

where θ and $V_{GS}-V_T$ denote the mobility-reduction coefficient and the effective bias of the transistor, respectively. Note that the channel-length modulation, body effect and the velocity saturation are neglected in (4.61) for the sake of analysis simplicity. The drain AC current can be described by the following one-dimensional power series as

$$i_d = g_m \cdot v_{gs} + K_{2_{gm}} \cdot v_{gs}^2 + K_{3_{gm}} \cdot v_{gs}^3 + \cdots \tag{4.62}$$

where g_m, $K_{2_{gm}}$ and $K_{3_{gm}}$ represent the linear, second-order and third-order coefficients of the transconductance gain of a transistor, respectively. The first three terms of the power series expansion are taken into account in the analyses for weak nonlinearities. The aforementioned coefficients of a transistor operating in saturation region can be obtained from (4.61) as follows:

$$\begin{aligned}
g_m &\equiv \frac{\partial I_D}{\partial V_{GS}} = \frac{\mu_o C_{ox}}{2}\frac{W}{L}(V_{GS}-V_T)\frac{1}{(1+\theta(V_{GS}-V_T))^2}(2+\theta(V_{GS}-V_T)) \\
K_{2_{gm}} &\equiv \frac{1}{2!}\frac{\partial^2 I_D}{\partial V_{GS}^2} = \frac{\mu_o C_{ox}}{2}\frac{W}{L}\frac{1}{(1+\theta(V_{GS}-V_T))^3} \\
K_{3_{gm}} &\equiv \frac{1}{3!}\frac{\partial^3 I_D}{\partial V_{GS}^3} = -\frac{\mu_o C_{ox}}{2}\frac{W}{L}\frac{\theta}{(1+\theta(V_{GS}-V_T))^4}.
\end{aligned} \tag{4.63}$$

The nonlinear output current generated by the input G_m stage of the filter is converted into a nonlinear voltage through the Q-boosted LC tank. Both the negative conductance generator and the varactor are assumed to be linear in this analysis where the contribution of the input G_m stage is sought. The detailed derivation of the 1dB compression point using the direct calculation of the nonlinear responses as an illustration of the method is given as an appendix at the end of the book. The analysis yields the following approximate expression for the input 1dB compression voltage

$$V_{1dB} \cong \sqrt{2.32 \times \frac{G_{m_{in}}^2 \left(1 + G_{m_{in}} R_{deg}\right)^3}{\left(2K_{2_{G_{m_{in}}}}^2 - K_{3_{G_{m_{in}}}} G_{m_{in}}\right)}}. \tag{4.64}$$

As (4.64) shows, the second and the third order nonlinearity coefficients should be reduced to increase the 1 dB compression voltage, and these require that the effective bias $V_{GS} - V_T$ of the input differential pair transistors be increased. An important difference between the expression obtained in the previous simplified analysis and (4.64) is the existence of the second-order nonlinearity term in the latter. The expression in (4.64) also demonstrates that it is possible to decrease the nonlinearity contribution by using source degeneration with extra power for the same input G_m. A very important observation to note is that as the nonlinearity due to velocity saturation and the nonlinearity of the output conductance of the transistors are neglected, the 1dB compression point for the input transconductance stage given in (4.64) does not show any dependancy to the filter quality factor.

The analysis for the nonlinearity contribution of the negative conductance generator using the method presented in [104] reveals a nonlinear term in the fundamental component of the output voltage causing compression in the power gain as the input amplitude increases. An expression for the nonlinear term at the fundamental frequency was obtained using a symbolic analysis tool, [107]. The tool implements the direct calculation of nonlinear responses method. Using the nonlinear term generated by the program in the derivation yields

$$V_{1dB} \cong \sqrt{\frac{2.32 \times g_{m5}}{G_m^2 \left(2K_{2_{g_{m5}}}^2 - K_{3_{g_{m5}}} g_{m5}\right)} \frac{1}{Q^3\, \omega_o^3\, L^3}} \tag{4.65}$$

where G_m and g_{m5} denote the input transconductance of the filter and the transconductance gain of one of the cross-coupled pair transistors, respectively. It is seen in

(4.65) that the second and the third order nonlinearity coefficients should be reduced to increase the 1 dB compression voltage. As (4.63) shows, the effective bias $V_{GS} - V_T$ of the cross-coupled negative conductance transistors should be increased to reduce the nonlinearity coefficients. Note that as the negative conductance generator is at the output of the filter, the higher the peak voltage gain at the center frequency, the larger the voltage swing across the negative conductance generator, thus the higher the contribution from the nonlinear cross-coupled pair decreasing the linearity. (4.65) links the 1dB compression point to the quality factor of the filter, Q, which is a measure of the selectivity. For a required center frequency, required filter quality factor Q and a fixed integrated tank, the required negative conductance $-g_{m_5}$ is a fixed value. Namely, $I_{SS}/(V_{GS} - V_T)$ of the cross-coupled transistors is fixed. Therefore, one should design the transistors in the cross-coupled pair to maximize their effective bias to improve the linearity, within the power consumption budget incorporated into the picture with I_{SS}. A simple design strategy for the cross-coupled negative conductance generator with a given LC tank for best linearity without taking into account the power consumption and noise constraints could be itemized as

- Determine the negative conductance $-g_m$ required for a specified filter Q.
- Maximize the effective bias of the transistors $V_{GS} - V_T$.
- Compute the corresponding tail current I_{SS}.

Fixing the I_{SS} at an upper limit for power consumption concern, one way to increase the effective bias for a certain transconductance value is to reduce the W/L of the transistors, as $g_m = \beta(V_{GS} - V_T)$ suggests.

The Q_o of the inductor has a considerable effect on the linearity. The higher the Q_o, the lower the loss conductance, G_{loss}, assumed to be dominated by the inductor.

For a fixed filter Q imposed by the selectivity specification of the targeted application, the lower the required transconductance of the negative conductance generating transistors. Thus, for the fixed tail current, I_{SS}, the dimensions of the cross-coupled transistors can be designed to allow for higher effective bias $V_{GS} - V_T$ for higher linearity.

Increasing the input transconductance and/or the quality factor of the filter increases the peak gain and this in turn degrades the linearity as evident in (4.65). A trade-off between the selectivity (the quality factor of the filter) and the linearity is observed. Note that as the negative conductance generator is at the output of the filter, the higher the peak voltage gain, the larger the voltage swing across the negative conductance generator, thus the higher the contribution from the nonlinear cross-coupled pair decreasing the linearity. Recall from (4.6) that the peak voltage gain of the filter can be expressed as

$$|H(j\omega_o)| \approx G_{m_{in}}\, Q\, \omega_o\, L \tag{4.66}$$

where $G_{m_{in}}$ is the input transconductance including the effect of the source degeneration. The input referred noise decreases with increasing input transconductance which reduces the noise contribution from the lossy tank and the negative conductance generator, pointing out to a strong trade-off between noise and linearity.

Adding source-degeneration resistors to the negative conductance generating cross-coupled transistors affect the linearity in the same way as it affects the linearity of the input transconductance stage. The input 1dB compression point is improved roughly by a factor of $\sqrt{(1 + g_{m_5} R_{deg})^3}$, where g_{m_5} and R_{deg} are the transconductance of the cross-coupled transistors and the associated source-degeneration resistor, respectively. The higher the source degeneration factor, the better the linearity, but the higher the power consumption to obtain a given filter quality factor with a fixed integrated tank.

The frequency tuning in the filter is provided by PMOS capacitors operated in depletion and inversion regions. The capacitance-voltage transfer characteristic of the PMOS capacitor (varactor) is highly nonlinear. The nonlinear capacitance-voltage transfer characteristic of the varactor is expanded to its power series around the value at $V_{control}$ up to the third-order nonlinearity coefficient. The varactor's region of operation is taken as depletion and the capacitance-voltage transfer characteristic is approximated by that of a nonlinear junction capacitor for simplicity, as it was elaborated in the previous subsection. The nonlinear term in the fundamental component that causes the compression was obtained using the symbolic analysis tool mentioned previously, [107]. A similar approach as the one used in the nonlinearity analysis of the negative conductance generator was followed in the derivation. Once again carrying out the analysis yields

$$V_{1dB} \cong \sqrt{\frac{1.55 \cdot \left|\frac{1}{j2\omega_o L} + G_{loss} + j2\omega_o\left(C + C_{Vo}\right)\right|}{Q^3\omega_o^3 L^3 G_m^2 \cdot \left|-\frac{K_{3_{C_V}}}{L} - j\omega_o 2K_{3_{C_V}} G_{loss} + 4\omega_o^2\left((C + C_{Vo})K_{3_{C_V}} - 2K_{2_{C_V}}^2\right)\right|}} \tag{4.67}$$

where $K_{2_{C_V}}$ and $K_{3_{C_V}}$ are the second and third-order nonlinearity coefficients in the power series expansion of the varactor's capacitance-voltage characteristic, respectively and $C_{Vo} = C_V(V_{control})$ is the quiescent value of the varactor capacitance. Note that as the varactor is at the output of the filter, the higher the peak voltage gain, $G_m\omega_o LQ$, the larger the voltage swing across the nonlinear varactor, thus the higher the contribution to the filter nonlinearity, similar to the case of the negative conductance generator. The inverse dependence of the linearity to the input transconductance is also obvious in (4.67), indicating the trade-off between noise and linearity, as previously mentioned. As the term Q^3 in the denominator of (4.67) shows, the higher the filter selectivity, namely the filter Q, the more pronounced the nonlinearity contribution from the varactor.

Finally, it should be noted that the derivations presented make the assumption of weak nonlinearity. The parasitic nonlinear junction capacitors, the nonlinearity of the output conductance of the transistors, the nonlinearity due to bulk-effect and the velocity saturation are not taken into account for the sake of obtaining simplified expressions. The nonlinearity simulations performed using SpectreRF show that the strongest contributors to the nonlinearity of the filter are the negative conductance generator and the varactor, respectively. The input 1dB compression points simulated with SpectreRF where only the components the nonlinearities of which are to be included are taken in transistor level, match with the corresponding figures calculated using (4.65) and (4.67) within less than 15% for a filter Q of 40. Comparisons between the simulated 1dB compression points of the filter with only the negative conductance generator and the varactor taken in transistor level and the corresponding theoretical 1dB compression points calculated using (4.65), and (4.67) as the filter Q is varied, are shown in Fig.68 and Fig.69, respectively. The calculations start to deviate from the simulations as the filter Q increases to values where the weak nonlinearity assumption which is the basis of the analyses presented above is no longer valid. The simplified analytical expressions presented prove useful in capturing the tendencies observed in the simulations as the circuit parameters are varied.

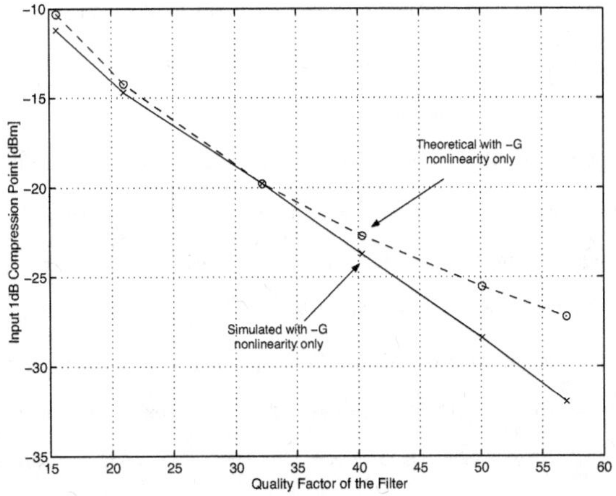

Fig. 68. Simulated and theoretical 1dB compression points of the filter with the negative conductance generator as the only nonlinearity contributor, versus the filter Q.

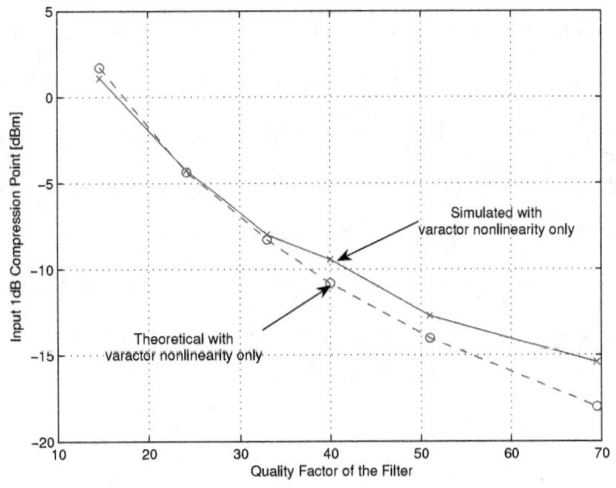

Fig. 69. Simulated and theoretical 1dB compression points of the filter with the varactor as the only nonlinearity contributor versus the filter Q.

2. Noise Considerations

This section is dedicated to the study of noise in the Q-enhanced filter. The aim is to obtain an interpretable expression for the noise figure yielding the trade-offs involved in the noise performance of the design. A similar analysis in the context of an LNA design with a notch filter is presented in [108]. The main noise contributors in the filter are the differential pair transistors and the source degeneration resistors in the input G_m stage, the cross-coupled transistors that generate negative conductance and the lossy tank elements. The losses in the integrated LC tank is dominated by the loss from the spiral inductor as mentioned before.

In order to accurately determine the noise contributed by the components in the filter, the negative conductance generated by the cross-coupled pair should be calculated somewhat accurately. The schematic of the cross-coupled NMOS differential pair that generates the differential negative conductance together with its small-signal equivalent circuit are shown in Fig.70. Note that the drain-source conductance g_{ds} of the transistors are also taken into account.

The circuit generates differential negative conductance through positive feedback. The value of the negative conductance is modified through the tail current, I_{SS}. From the equivalent small signal circuit that is also depicted in Fig.70, the following can be written for the branch currents

$$\begin{aligned} i &- g_m v_2 - g_{ds} v_1 = 0 \\ i &+ g_m v_1 + g_{ds} v_2 = 0 \end{aligned} \tag{4.68}$$

where g_m is the transconductance and g_{ds} is the drain-source conductance of the cross-coupled transistors, which are assumed to be identical. Note that the node

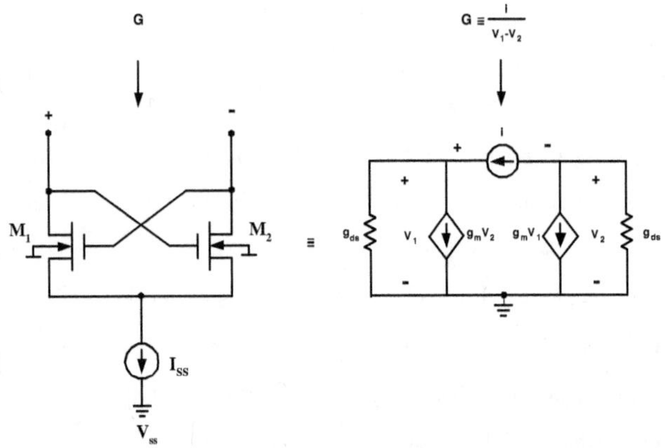

Fig. 70. Cross-coupled differential pair as negative conductance generator and its small-signal equivalent circuit.

where the source terminals of the transistors are coupled is taken as ground because of the differential excitation of the circuit. Adding the two equations in (4.68) and solving for i yields

$$i = \frac{-g_m(v_1 - v_2) + g_{ds}(v_1 - v_2)}{2}. \tag{4.69}$$

Using the definition of the differential conductance as given in Fig.70 together with (4.69), the following is obtained

$$G \equiv \frac{i}{v_1 - v_2} = -\frac{(g_m - g_{ds})}{2}. \tag{4.70}$$

As the derivation above shows, a cross-coupled differential pair generates a negative differential conductance with a value directly proportional to the transconductance of each transistor. The drain-source conductance of the transistors on the other hand reduces the amount of the negative conductance. The lower the quality factor of the integrated LC tank in the filter, the higher the loss associated with it and the larger the value of the negative conductance for a given filter quality factor. One should simply increase the tail current I_{SS} of the cross-coupled differential pair to increase the value of the negative conductance.

A small-signal equivalent circuit of the single-ended version of the RF filter used as a basis for the noise analysis is shown in Fig.71, where the negative conductance generator is represented as a negative conductance of $-G$ together with its mean square noise current denoted as $\overline{I_{-G}^2}$.

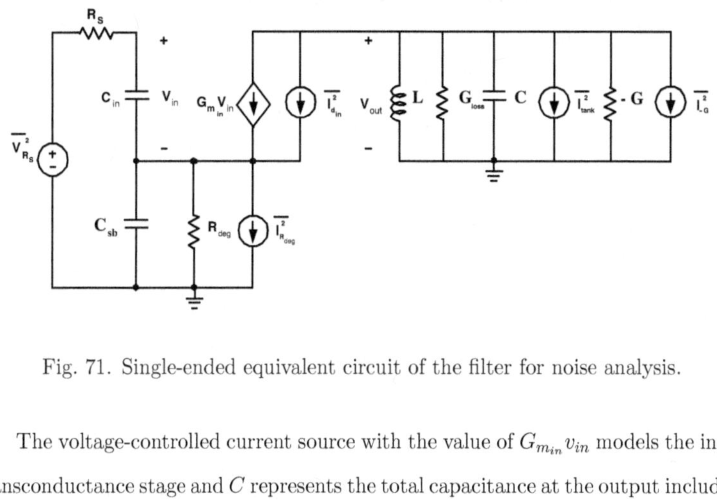

Fig. 71. Single-ended equivalent circuit of the filter for noise analysis.

The voltage-controlled current source with the value of $G_{m_{in}} v_{in}$ models the input transconductance stage and C represents the total capacitance at the output including the varactor capacitance. G_{loss} in the equivalent circuit models the loss conductance of the tank, which is dominated by the loss from the spiral inductor. R_S denotes the source resistance and $\overline{V_{R_S}^2}$ is the mean square noise voltage of the source resistance. R_{deg} is the source degeneration resistor of the input G_m transistor (input differential pair transistor) and $\overline{I_{R_{deg}}^2}$ denotes its mean square noise current. The drain noise current power of the input G_m transistor is $\overline{I_{d_{in}}^2}$, whereas the noise current power of the LC tank is denoted as $\overline{I_{tank}^2}$. The input capacitance of the filter is represented as C_{in}, whereas C_{sb} models the source-bulk junction capacitance of the input transistor. The drain-source conductance g_{ds} and the bulk transconductance g_{mbs} of the input transistor is neglected in Fig.71 for the sake of clarity in the equivalent circuit and simplicity in the analysis to follow. Although the g_{ds} and g_{mbs} of the input transistor are taken into account in the MATLAB computations based on the analytical model presented below and compared with the simulated characteristics in the following figures, it should be noted that their inclusion in the derivations adds a negligible accuracy to the derived formula. The magnitude of the voltage gain of the filter can be expressed using the model in Fig.71 as

$$|H(j\omega)| \cong \left| -\frac{G_{m_{in}} / \left(1 + G_{m_{in}} Z_{deg}(j\omega)\right)}{G_{loss} - G + j\omega C - j\frac{1}{\omega L}} \right| \tag{4.71}$$

where $Z_{deg}(j\omega)$ is the source degeneration impedance which is the parallel combination of R_{deg} and C_{sb}. Note that the attenuation due to the input capacitance of $C_{in} \cong 120fF$ (the value would somehow be decreased due to the degeneration) is neglected. The frequency characteristic of the magnitude of the filter's voltage gain as calculated using (4.71) for a filter quality factor of $Q = 8$ is plotted in Fig.72 together with the simulated characteristic.

As the aim of this noise analysis is to derive an expression for the noise figure of the filter at the center frequency $\omega_o \cong 1/\sqrt{LC}$, the noise contribution of each component at the center frequency should be determined. The total output noise can then be computed through superposition. It should be noted that the contributions from the input differential pair transistor, the cross-coupled transistor, inductor and the source-degeneration resistor should be multiplied by two as the filter is differential. It is clearly seen in Fig.71 that in calculating the output noise contribution of the

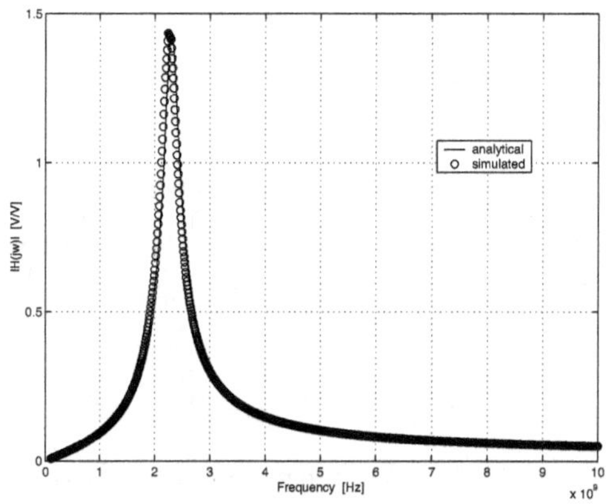

Fig. 72. Frequency characteristic of the magnitude of the filter's voltage gain with Q=8.

source resistance R_S at ω_o, its mean square noise voltage is to be multiplied by the square of the magnitude of the voltage gain at ω_o:

$$\left.\overline{V^2_{out_{R_S}}}\right|_{\omega=\omega_o} = \overline{V^2_{R_S}} \cdot |H(j\omega_o)|^2 \cong 4kTR_S \times \left(\frac{G_{m_{in}}}{|1 + G_{m_{in}} Z_{deg}(j\omega_o)|\,(G_{loss} - G)}\right)^2 \quad (4.72)$$

where k and T denote the Boltzmann's constant and the temperature in Kelvin, respectively. The drain noise current power spectral density of an MOS transistor is calculated using the following formula

$$\frac{\overline{I_d^2}}{\Delta f} = 4kT\frac{2}{3}\left(g_m + g_{mbs} + g_{ds}\right) \quad (4.73)$$

where g_m, g_{mbs} and g_{ds} denote the transconductance gain, the bulk transconductance and the drain-source conductance of an MOS transistor, respectively, [72]. The drain noise current power spectral densities of the cross-coupled transistors and the input

G_m stage differential pair transistors are calculated using (4.73). It should be noted that the mean square drain noise current of the cross-coupled transistor is denoted as $\overline{I_{-G}^2}$ in Fig.71. The output noise contribution of the input G_m transistor at ω_o can be computed by first referring the mean square drain noise current denoted as $\overline{I_{d_{in}}^2}$ in Fig.71 to the input of the filter and then multiplying the resultant mean square noise voltage with the square of the magnitude of the voltage gain at ω_o.

$$\overline{V_{out_{d_{in}}}^2}\bigg|_{\omega=\omega_o} = \frac{\overline{I_{d_{in}}^2}}{G_{m_{in}}^2} \cdot |H(j\omega_o)|^2. \tag{4.74}$$

Observe in Fig.71 that, the output noise contributions of the tank (i.e. the inductor) and the negative conductance generator can be calculated by multiplying their mean square noise currents with the square of the output impedance of the filter at the resonance.

$$\begin{aligned}
\overline{V_{out_{tank}}^2}\bigg|_{\omega=\omega_o} &= \overline{I_{tank}^2} \cdot |Z_{out}(j\omega_o)|^2 = \overline{I_{tank}^2} \cdot \left(\frac{1}{G_{loss} - G}\right)^2 \\
\overline{V_{out_{-G}}^2}\bigg|_{\omega=\omega_o} &= \overline{I_{-G}^2} \cdot |Z_{out}(j\omega_o)|^2 = \overline{I_{-G}^2} \cdot \left(\frac{1}{G_{loss} - G}\right)^2.
\end{aligned} \tag{4.75}$$

The noise current power spectral densities of the cross-coupled transistor, input G_m stage differential pair transistor, the source degeneration resistor and the LC tank are given in (4.76) for the sake of completeness

$$\begin{aligned}
\frac{\overline{I_{-G}^2}}{\Delta f} &= 4kT\frac{2}{3}\left(g_{m_{cc}} + g_{mbs_{cc}} + g_{ds_{cc}}\right) \\
\frac{\overline{I_{d_{in}}^2}}{\Delta f} &= 4kT\frac{2}{3}\left(G_{m_{in}} + g_{mbs_{in}} + g_{ds_{in}}\right) \\
\frac{\overline{I_{R_{deg}}^2}}{\Delta f} &= 4kT\frac{1}{R_{deg}}
\end{aligned}$$

$$\frac{\overline{I_{tank}^2}}{\Delta f} = 4kTG_{loss} \tag{4.76}$$

where $g_{mbs_{in}}$, $g_{ds_{in}}$ and $g_{mbs_{cc}}$, $g_{ds_{cc}}$ denote the bulk transconductance and the drain-source conductance of the input differential pair transistor and the cross-coupled transistor, respectively. The output noise contribution of the source degeneration resistor at the resonance can be calculated by first converting its mean square noise current into mean square noise voltage. This is done by multiplying the noise current power with the square of magnitude of the equivalent impedance at the source node of the input transistor at the center frequency. The resulting mean square noise voltage multiplied with the square of the magnitude of the common gate voltage gain of the input transistor at the center frequency gives the output noise contribution of the source degeneration resistor.

$$\left.\overline{V_{out R_{deg}}^2}\right|_{\omega=\omega_o} \cong \overline{I_{R_{deg}}^2} \cdot |Z_{s_{in}}(j\omega_o)|^2 \cdot \left(\frac{G_{m_{in}}}{|Y_{out}(j\omega_o)|}\right)^2 \tag{4.77}$$

where $|Y_{out}(j\omega_o)|$ denotes the magnitude of the output admittance of the filter at resonance and $|Z_{s_{in}}(j\omega_o)|$ is the magnitude of the equivalent impedance at the source node of the input transistor at the center frequency given as

$$|Z_{s_{in}}(j\omega_o)| \cong \left|\frac{1}{G_{m_{in}} + \frac{1}{R_{deg}} + j\omega_o\left(C_{in} + C_{sb}\right)}\right| . \tag{4.78}$$

Having determined the contributions from the noise generating components in the filter, the total output noise power spectral density can be expressed using superposition.

$$\frac{\overline{V_{out}^2}}{\Delta f} = 2\frac{\overline{V_{out_{tank}}^2}}{\Delta f} + 2\frac{\overline{V_{out-G}^2}}{\Delta f} + 2\frac{\overline{V_{out R_{deg}}^2}}{\Delta f} + 2\frac{\overline{V_{out_{d_{in}}}^2}}{\Delta f} + \frac{\overline{V_{out R_S}^2}}{\Delta f}. \tag{4.79}$$

It should be emphasized that in (4.79), all the power spectral densities are func-

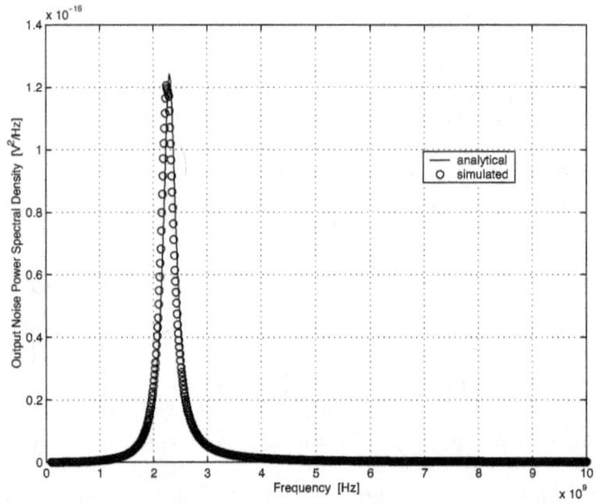

Fig. 73. Output noise power spectral density of the filter with Q=8.

tions of the frequency as they are shaped by the bandpass characteristic of the filter. The output noise power spectral density as calculated from (4.79) is plotted in Fig.73 together with the simulated noise of the transistor-level schematic. The parameters of the filter are taken such that the filter quality factor is 8 at 2.24GHz.

Noise figure of a circuit provides a measure of how much the signal-to-noise ratio (SNR) is degraded from the input to the output of the circuit. The most commonly accepted definition of noise figure NF is

$$NF \equiv 10 \log_{10} \left(\frac{SNR_{in}}{SNR_{out}} \right) \tag{4.80}$$

where SNR_{in} and SNR_{out} denote the signal-to-noise ratios at the input and the output of the circuit, [7]. The argument of the logarithm in (4.80) is the *Noise Factor* and it is denoted as F. In a noiseless circuit, as the signal-to-noise ratio is not altered from the input to the output (the circuit does not add any noise), the noise figure NF is equal to 0dB. Obviously, as all the circuits in reality generate some noise, the

SNR at the output is lower than the SNR at the input, yielding $NF > 0dB$, $(F > 1)$. Noise figure is usually specified in a unit bandwidth at a given frequency of interest and it is also called "spot" noise figure, [7]. Substituting $SNR = S/N$ where S and N denote the signal and noise power, respectively, into (4.80) and rearranging the latter gives

$$NF = 10\log_{10}\left(\frac{S_{in}}{S_{out}} \cdot \frac{N_{out}}{N_{in}}\right) = 10\log_{10}\left(\frac{1}{|H(j\omega)|^2} \cdot \frac{\overline{V_{out}^2}}{4ktR_S}\right). \tag{4.81}$$

Note that the SNR_{in} is assumed to be the ratio of the input signal power to the noise generated by the source resistance R_S, modelled as $\overline{V_{R_S}^2} = 4kTR_S$. As mentioned at the beginning of this section, the aim of the noise analysis presented here is to obtain an approximate interpretable expression for the noise figure at the center frequency of the filter. Once the mean square output noise voltage at the center frequency is determined in accordance with (4.79) and substituted into (4.81) for $\overline{V_{out}^2}$, then it takes replacing $|H(j\omega)|$ with the magnitude of the voltage gain at the center frequency (i.e. the peak gain) $|H(j\omega_o)|$ to obtain an expression for the noise figure at the center frequency. For the sake of clarity, the noise factor F will be derived from this point on, as the noise figure is simply $NF = 10\log_{10}(F)$.

The noise factor at the center frequency can be obtained from (4.81) as

$$F\big|_{\omega=\omega_o} = F_{\omega_o} = \frac{1}{|H(j\omega_o)|^2} \cdot \frac{\overline{V_{out}^2}\Big|_{\omega=\omega_o}}{4ktR_S}. \tag{4.82}$$

Substituting (4.79) into (4.82) yields

$$F_{\omega_o} = \frac{\left(2\overline{V_{out_{tank}}^2} + 2\overline{V_{out-G}^2} + 2\overline{V_{out_{R_{deg}}}^2} + 2\overline{V_{out_{d_{in}}}^2} + \overline{V_{out_{R_S}}^2}\right)}{|H(j\omega_o)|^2 4ktR_S} \tag{4.83}$$

where all the mean square noise voltages at the numerator are to be evaluated at the

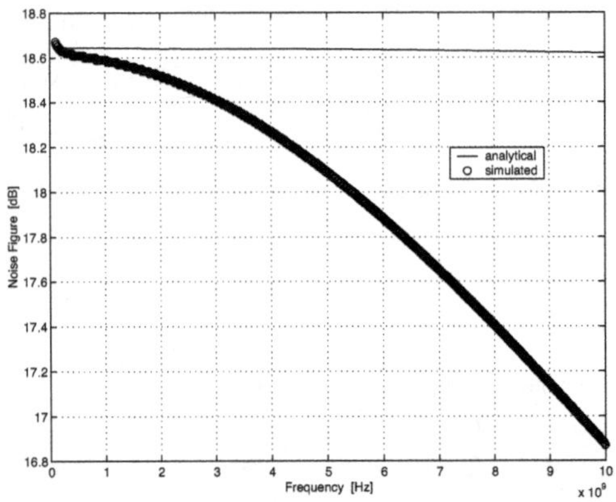

Fig. 74. Noise figure of the filter with Q=8.

center frequency of the filter. Note that the mean square noise voltage contributions of all the components to the output noise were determined previously. Substituting the associated expressions into (4.83) and rearranging the latter produces

$$F_{\omega_o} = \frac{\left(\left(\frac{8kTG_{loss}+2\overline{I^2_{-G}}+2\overline{I^2_{R_{deg}}}G^2_{m_{in}}\left|Z_{s_{in}}(j\omega_o)\right|^2}{(G_{loss}-G)^2}\right)+|H(j\omega_o)|^2\left(\frac{2\overline{I^2_{d_{in}}}}{G^2_{m_{in}}}+4ktR_S\right)\right)}{|H(j\omega_o)|^2 4ktR_S} \tag{4.84}$$

where $\overline{I^2_{-G}}$, $\overline{I^2_{d_{in}}}$ and $\overline{I^2_{R_{deg}}}$ are given in (4.76). The noise figure calculated using (4.81) and (4.79) is plotted in Fig.74 together with the simulated noise figure of the transistor-level schematic. The simulated noise figure drops considerably faster than the calculated one does as the frequency increases but it should be noted that the x-axis extends to 10 GHz, which is well beyond the frequency of interest. The difference between the calculated and simulated values at the center frequency of $f_o \cong 2.3$ GHz is less than 0.2 dB. Table VI compares the simulated and calculated noise related characteristics of the filter and tabulates the associated relative error at each characteristic for a filter Q of 8. Note that the relative error of the noise factor at resonance is less than 5% as shown in Table VI.

Table VI. Simulated vs. calculated noise related characteristics of the filter (Q=8).

Filter Characteristic	Simulated	Calculated	Relative Error
Peak Gain [V/V]	1.44	1.43	-0.7 %
Center Frequency f_o [GHz]	2.24	2.3	2.7 %
Maximum Output Noise [V^2/Hz]	120.6×10^{-18}	124.2×10^{-18}	3 %
Inductor Noise Contribution	49.2 %	48.8 %	-0.81 %
Negative Conductance Noise Contribution	35.1 %	34.76 %	-0.86 %
Input G_m Stage Noise Contribution	12.58 %	12.43 %	-1.2 %
Source Degeneration Noise Contribution	2.72 %	2.62 %	-3.68 %
Noise Figure at f_o [dB]	18.5	18.64	3.3 %

As the quality factor of the filter and thus the peak gain of the filter is increased, the error that the approximations in the analysis introduce may be more pronounced. In order to test whether the noise analysis still provides a reasonable accuracy for the noise figure at a higher filter Q, the simulated and calculated filter characteristics for a filter Q of 20 are provided in Figs.75,76 and 77.

It is seen in Fig.75 that as the calculated magnitude of the voltage gain is around 10% higher than the simulated value at their respective center frequencies, the calculated maximum output noise voltage power density also comes out to be higher than the simulated value, shown in Fig.76. This is observed to be due mainly to the discrepancy between the simulated and calculated contributions of the source resis-

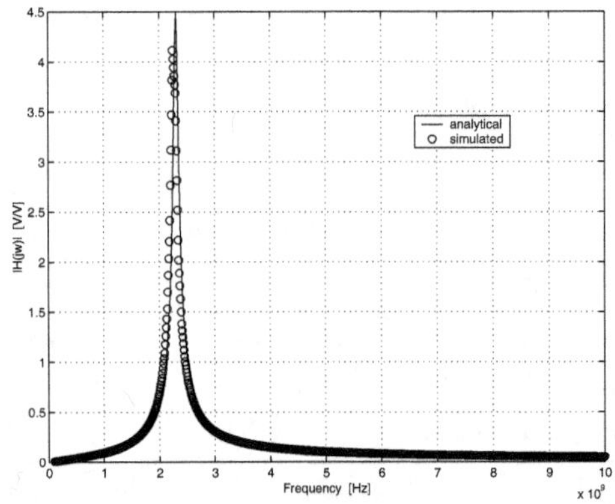

Fig. 75. Frequency characteristic of the magnitude of the voltage gain of the filter with Q=20.

tance noise. Similar to the previous case, the simulated noise figure in Fig.77 drops considerably faster than the calculated one does as the frequency increases but once again it should be noted that the x-axis extends to 10 GHz, which is well beyond the frequency of interest. The difference between the calculated and simulated values at the center frequency of $f_o \cong 2.3$ GHz is less than 0.2 dB.

Table VII compares the simulated and calculated noise related characteristics of the filter and tabulates the associated relative error at each characteristic for a filter Q of 20. Note that although the relative errors at some of the other noise characteristics are somewhat increased with respect to those obtained for filter Q of 8 due to the more pronounced effect of the source resistance noise at the output as the gain is increased, the error at the noise factor at resonance is less than 5% as shown in Table VII. Recall that the discrepancy between the simulated and calculated contributions of the source resistance noise at the output is not reflected in the noise figure values as much as it is reflected in the maximum output noise densities since there is a kind

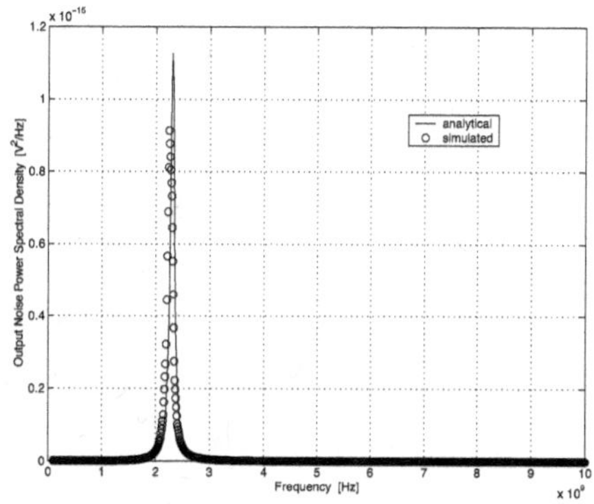

Fig. 76. Output noise power spectral density of the filter with Q=20.

of normalization with respect to the source resistance noise at the output in noise figure calculation as it is seen in (4.81).

The expression for the noise factor at the resonance given in (4.84) can be rearranged in order for the trade-offs to be better visualized as below

$$F = 1 + \frac{8kTG_{loss} + 2\overline{I^2_{-G}} + 2\overline{I^2_{R_{deg}}}G^2_{m_{in}}\left|Z_{s_{in}}(j\omega_o)\right|^2}{4kTR_S\acute{G}^2_{m_{in}}} + \frac{2\overline{I^2_{d_{in}}}}{4kTR_SG^2_{m_{in}}} \tag{4.85}$$

where $\acute{G}_{m_{in}}$ denotes the input transconductance attenuated by the source degeneration and it is given as

$$\acute{G}_{m_{in}} = \frac{G_{m_{in}}}{\left|1 + G_{m_{in}}Z_{deg}(j\omega_o)\right|}. \tag{4.86}$$

Note that the attenuated input transconductance is evaluated at the center frequency, as the noise factor of interest is that at the center frequency. A small value of 30Ω was used in the RF filter as source degeneration resistor in order to keep the its negative impact on the noise factor low while providing some improvement on the linearity

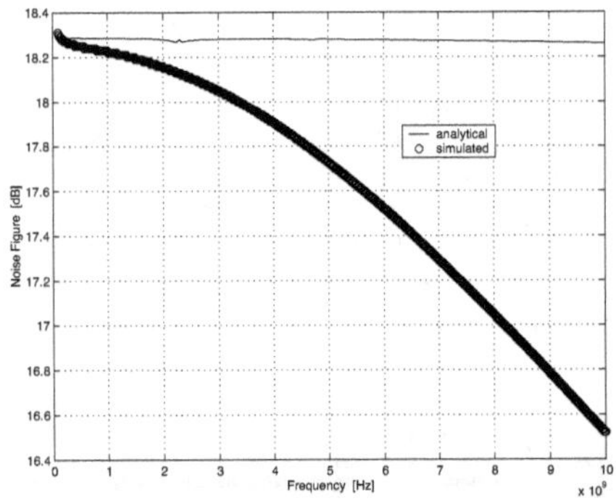

Fig. 77. Noise figure of the filter with Q=20.

performance of the filter. As it is observed in (4.85), increasing the value of the source degeneration resistor and at the same time keeping $G_{m_{in}}$ constant reduces the input transconductance $\acute{G}_{m_{in}}$, thus increasing the contributions of the negative conductance generator and the LC tank, which are already the dominant noise contributors as evident in Tables VI and VII. In case the input transconductance is to be kept constant in the design, then the increase at the source degeneration resistor value should be accompanied by an increase at the transconductance gain of the input transistor with more power consumption (this can be done if $G_{m_{in}} R_{deg} \gg 1$ is not valid). Besides, as long as the value of the source degeneration resistor is dominantly determining the magnitude of the impedance at the source node of the input transistor at the resonance, $|Z_{s_{in}}(j\omega_o)|$, increasing its value also increases its noise contribution to the noise factor expressed in (4.85). This discussion shows that the source degeneration resistor used to improve the linearity of the input G_m stage either degrades the noise performance of the filter yielding a trade-off between linearity and noise, or requires more power consumption for this trade-off to be avoided.

Table VII. Simulated vs. calculated noise related characteristics of the filter (Q=20).

Filter Characteristic	Simulated	Calculated	Relative Error
Peak Gain [V/V]	4.13	4.5	9 %
Center Frequency f_o [GHz]	2.24	2.3	2.7 %
Maximum Output Noise [V^2/Hz]	917×10^{-18}	1126×10^{-18}	22.8 %
Inductor Noise Contribution	44.8 %	44.46 %	-0.76 %
Negative Conductance Noise Contribution	38.1 %	37.8 %	-0.8 %
Input G_m Stage Noise Contribution	13.66 %	13.52 %	-1 %
Source Degeneration Noise Contribution	2.96 %	2.74 %	-7.4 %
Noise Figure at f_o [dB]	18.13	18.28	3.5 %

An even more important trade-off between linearity and noise is imposed at the input transconductance design. Recall from the nonlinearity analysis presented in the previous section that increasing the input transconductance increases the peak gain and this in turn degrades the linearity of the filter considerably as the nonlinearities of the negative conductance generator and the varactor are more pronounced with the increased voltage swing across them. On the other hand, increasing the input transconductance clearly decreases the contributions of the input G_m transistor, the negative conductance generator and the LC tank to the noise factor as evident in (4.85), improving the noise performance of the filter. Taking the dynamic range as the main performance metric, it is observed that reducing the input transconductance and consequently the peak gain through lowering the tail current of the input differential pair G_m stage helps to considerably reduce the power consumption at the same time keeping the dynamic range almost constant. This is because the noise figure is degraded by almost the same amount the 1dB compression point of the filter increases to keep the dynamic range nearly constant. The trade-off is demonstrated in the section where the measurement results of the RF filter is presented.

The first term in the numerator of the second term at the right-hand side of (4.85) models the contribution of the LC tank the loss of which is dominated by that of the inductor. Note that as the quality factor of the inductor increases, its loss conductance G_{loss} decreases, and consequently its contribution to the noise factor decreases, improving the noise performance of the filter. The decrease in the loss conductance of the tank (i.e. inductor) obviously increases the filter quality factor given that the negative conductance is kept constant. This is how the filter quality factor is increased from around 8 to 20 in the comparison tests presented in the figures and tables above. The decrease at the noise figure given in Table VII for $Q = 20$ with respect to the noise figure given in Table VI for $Q = 8$ verifies the argument that as the inductor quality factor increases, the noise figure decreases.

The second term in the numerator of the second term at the right-hand side of (4.85) models the contribution of the negative conductance generating cross-coupled pair. As it is seen in (4.76) and (4.70), increasing the tail current of the cross-coupled negative conductance generator to increase the negative conductance for a higher filter Q makes the associated mean square noise current $\overline{I_{-G}^2}$ larger, consequently increasing the noise factor. Note that as the quality factor of the inductor increases, a lower negative conductance $-G$ is required from the cross-coupled pair for a given filter Q dictated by the selectivity requirements of the system the filter is designed for. This in turn decreases the mean square noise current generated by the negative conductance generator, reducing the noise factor further in addition to the reduced noise contribution from the inductor. Another obvious advantage of using an inductor with a higher quality factor is a reduction in the power consumption as the required negative conductance and thus the required tail current is decreased.

The noise factor expression given in (4.85) accurately captures the tendencies observed in the simulations as the circuit parameters are varied. It should be noted that, the noise simulations of the filter are also performed using the Periodic Steady-State (PSS) noise analysis of SpectreRF to take into account the nonlinear time-varying effects. The PSS noise simulation predicts a noise figure of 27dB with a filter Q of 40 at around 2.27GHz whereas the linear noise simulation yields a noise figure of 19dB. The measured noise figure for a filter Q of 40 at 2.19GHz is 26.8dB, matching very well with the PSS noise simulation result.

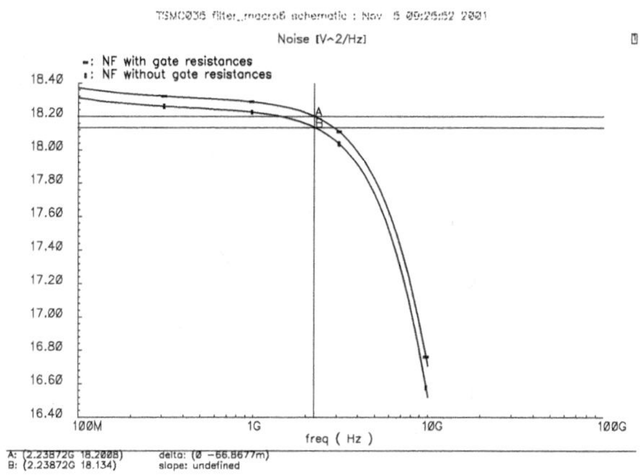

Fig. 78. Effect of the gate resistances on the noise figure of the filter with Q=20.

The thermal noise contribution of the gate resistances of the transistors in the filter is neglected in the noise analysis carried out above. The transistors in the filter were laid out in multiple fingers to reduce their gate resistances. The gate resistance of a transistor with N fingers of W/L dimensions each, can be computed as

$$R_{gate} = \frac{1}{3}\frac{W}{L}\frac{1}{N}R_{\Box_{poly}} \tag{4.87}$$

where $R_{\Box_{poly}}$ denotes the sheet resistance of gate poly per square and $1/3$ on the right hand side accounts for the distributed nature of the resistance, [109]. The use of multiple fingers in the transistors (10 fingers at the input G_m differential pair transistors and 12 fingers at the cross-coupled negative conductance generating transistors) reduces the associated gate resistances computed using (4.87) to values less than 5Ω. Adding the gate resistances explicitly into the circuit schematic to better assess their noise contribution shows an increase of merely around 0.07dB in the simulated noise figure at the center frequency. The simulation results with and without the gate resistances are given in Fig.78.

3. Dynamic Range Simulations

The most important aspect of an RF bandpass filter is its dynamic range. The following section demonstrates how the dynamic range of the filter is computed based on the simulations. Dynamic range is limited at low power levels by the noise level. The minimum detectable signal power level at the input is taken as a level 3dB higher than the thermal noise floor within the noise bandwidth, [110]. The thermal noise within a unit bandwidth in dBm is calculated as follows

$$10\log\left(\frac{kT}{1mW}\right) = 10\log\left(\frac{1.38\times10^{-23}\times290}{1\times10^{-3}}\right) \cong -174dBm/Hz, \tag{4.88}$$

where k and T denote the Boltzmann's constant and the temperature in kelvin, respectively. Thus, the minimum detectable signal power level at the input, which limits the dynamic range at low-end can be computed as

$$P_{i,mds} = -174dBm + 10\log B + F(dB) + 3dB, \tag{4.89}$$

where F denotes the noise figure of the filter at the center frequency and B is the bandwidth over which the noise is integrated, typically taken as the equivalent noise bandwidth of the filter. With V_{dd}=1.8V, I_{Vdd}=6mA, Q=8.5, f_o=1.7GHz, the 1 dB compression point and the noise figure are simulated to be around -13.5dBm and 14.2dB in SpectreRF. The in-band dynamic range is defined as the ratio of the 1dB compression point input power to the minimum detectable signal at the input with the noise integrated over the equivalent noise bandwidth. Note that the bandwidth of the filter is $\Delta f = f_o/Q = 1.7GHz/8.5 \cong 200MHz$ in the simulations with the bias settings given above. The equivalent noise bandwidth of a second-order transfer function is approximated with $\pi/2\times\Delta f$, [43]. Therefore, the equivalent noise bandwidth of the second-order filter would be $B = 200MHz \times \pi/2 \cong 314.16MHz$. Substituting the equivalent noise bandwidth and the noise figure values into (4.89) yields

$$P_{i,mds} = -174dBm/Hz + 10\log(314.16 \times 10^6) + 14.2 + 3dB \cong -71.83dBm. \quad (4.90)$$

The in-band dynamic range, also called as 1dB compression dynamic range, can now be calculated as

$$DR_{1dB} = P_{i,1dB} - P_{i,mds} = -13.5 - (-71.83) = 58.33dB. \quad (4.91)$$

The third-order intercept point at the input, IIP_3, is simulated to be around -3.8dBm for the given bias settings. This figure is used to calculate the spurious-free dynamic range, $SFDR$. The IIP_3 determines the highest allowable power level at the input of the filter for a spurious-free output spectrum as it can be related analytically to the power level that generates an output third-order component equal to the noise floor, [110].

$$SFDR = \frac{2}{3}(IIP_3 - P_{i,mds}) = \frac{2}{3}(-3.8 - (-71.83)) = 45.35dB. \quad (4.92)$$

Note that with the Q setting of 8.5, the filter bandwidth is quite large and the attenuation in the stopband is poor. In the simulations, the amount of positive feedback is increased to boost the Q of the filter for two settings. With the first setting of V_{dd}=1.5V, I_{Vdd}=7.7mA, Q=29, f_o=2.228GHz, the 1 dB compression point and the noise figure are simulated to be around -29.6dBm and 14.8dB in SpectreRF. The 3dB bandwidth is simulated as 77MHz, which yields an equivalent noise bandwidth of 121MHz. With the second setting of V_{dd}=1.5V, I_{Vdd}=7.8mA, Q=42, f_o=2.228GHz,

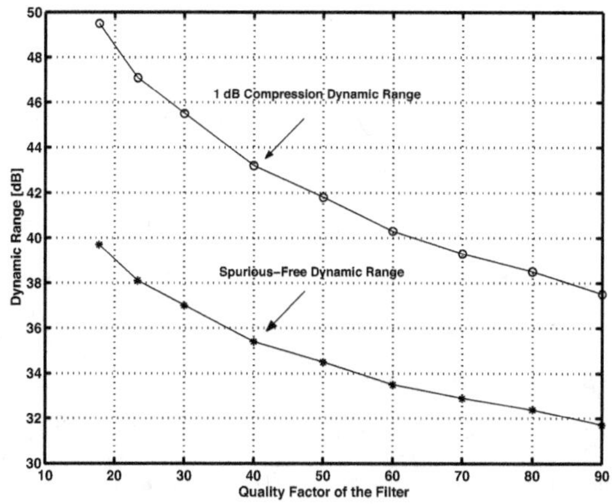

Fig. 79. Simulated 1dB compression point and spurious-free dynamic range versus the quality factor of the filter.

the 1 dB compression point and the noise figure are simulated to be around -33.9dBm and 14.9dB in SpectreRF. The 3dB bandwidth in this case is 53MHz, yielding an equivalent noise bandwidth of 83.2MHz. The two-tone intermodulation distortion simulations with SpectreRF require unacceptably large memory as the inputs need to be relatively close to each other at these high Q values. Therefore, input third-order intercept points are taken as roughly 10dB larger than the associated 1 dB compression points. Substituting the corresponding simulated values into (4.89), the minimum detectable signal levels are computed as -75.4dBm and -76.9dBm, for the first and the second settings, respectively. The 1dB compression and spurious-free dynamic range figures can easily be computed using (4.91) and (4.92). The substitutions for the first setting yield a 1dB compression dynamic range of 46dB and an SFDR of 37dB, whereas the substitutions for the second setting yield a 1dB compression dynamic range of 43dB and an SFDR of 35dB. Note that the filter

Q is inversely proportional to the dynamic range achievable as suggested by the approximate expression given in (4.2).

The three main contributors to the nonlinearity of the filter are the negative conductance generator, the varactor and the input g_m stage. The nonlinearity analysis of the circuit demonstrates that the contributions of the negative conductance generator and the varactor are much more pronounced than that of the input stage as the former two components across their inputs are subject to the full voltage swing at the output of the filter. As the filter Q increases, the voltage swing at the output increases because of the higher peak gain given by (4.6), and so do the nonlinearity contributions from the negative conductance circuit and the varactor. One possible way to improve the dynamic range is to linearize the negative conductance circuit with source-degeneration at the expense of more power consumption. The strong dependance of the filter dynamic range to the filter Q, as (4.2) suggests is illustrated in Fig.79.

CHAPTER V

INTEGRATED CIRCUIT MEASUREMENTS

This chapter includes the measurement results of the integrated circuits (ICs) designed throughout the research. All the prototype circuits were fabricated through MOSIS with either HP 0.5μm or TSMC 0.35μm CMOS technologies with the exception of the RF Prescaler fabricated with 0.6μm BiCMOS Technology of Texas Instruments. The measurement results of each integrated circuit is evaluated in the context of a wireless communication standard specification in order to determine the bottlenecks in the performance of the circuits and provide some design guidelines for future improvements.

The laboratory setups used in characterizing the ICs are shown in Figs.80 and 81.

Fig. 80. Test setup in the laboratory used for the characterization of the ICs.

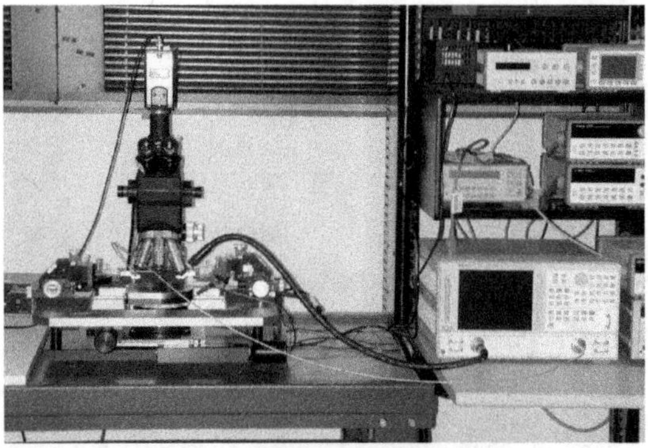

Fig. 81. Test setup in the laboratory used for the on-wafer characterization of the ICs.

A. Negative-Conductance LC oscillators in CMOS

1. Fully-Integrated CMOS LC VCOs with NMOS Cross-Coupled Pairs

Two CMOS VCOs using only NMOS cross coupled pairs for negative-conductance generation to compensate for the LC tank losses were designed and fabricated with HP 0.5μm and TSMC 0.35μm CMOS technologies through MOSIS. The chip micrograph of the VCO with HP 0.5μm CMOS is shown in Fig.82. Both VCOs were bonded to TQFP 44 packages and the measurements were taken on the packaged parts. The spiral inductors used in the VCOs are identical in geometry to the structure mentioned in Chapter II.

The IC that includes the NMOS VCO designed with HP 0.5μm CMOS technology also accomodates a CMOS VCO with complementary cross-coupled differential pairs. Two different PCBs were designed to characterize these VCOs. 30 ICs were received from fabrication and in order to be able to test the functionality of all the VCOs, a PCB was designed around a 40 pin ZIF (Zero-Insertion-Force) socket. The PCB is shown in Fig.83.

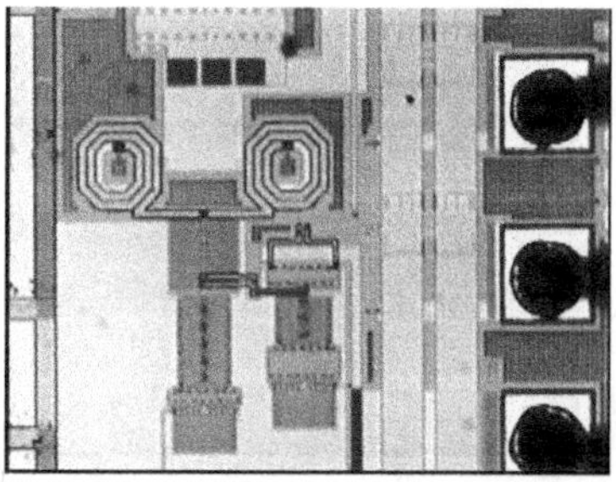

Fig. 82. Die micrograph (0.28mm X 0.33mm) of the VCO with HP 0.5μm CMOS.

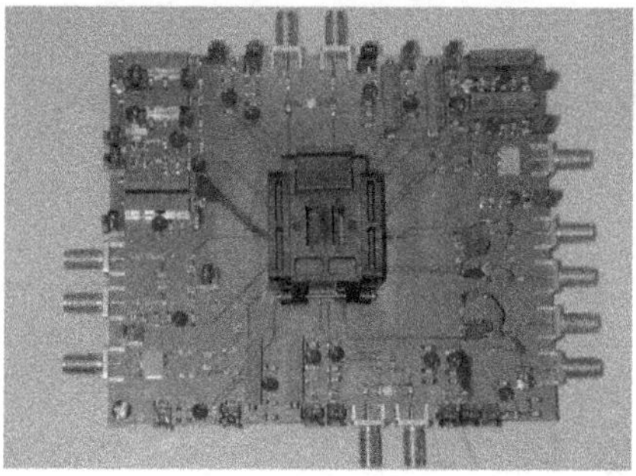

Fig. 83. PCB designed to test packaged VCOs with HP 0.5μm CMOS.

After all the ICs were tested for the oscillation frequency and the fundamental output power, one of the ICs that demonstrated a relatively good performance was soldered on the PCB designed for noise and frequency tuning characterization with the intention of minimizing the impact of the package parasitics on the measured performance. This second PCB is shown in Fig.84. The design of the PCB was done carefully to minimize the board parasitics.

Resistors and bypass capacitors were placed on all the bias and supply lines on both PCBs for noise filtering purposes. A schematic representation of the RC

Fig. 84. PCB designed for RF characterization of the packaged VCOs with HP 0.5μm CMOS.

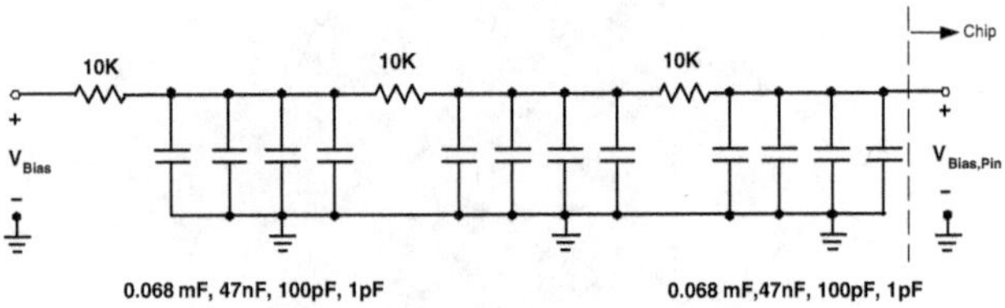

Fig. 85. Schematic representation of the filtering on the bias and supply lines on the PCBs.

filtering sections used is depicted in Fig.85. The filtering with three sections as illustrated was done on the sensitive frequency control voltage lines of the CMOS and NMOS VCOs. Four capacitors with different values ranging from 68μF to 1pF were used in each section in order to improve the efficiency of the filtering. Note that a 1pF capacitor would provide a better bypass path to noise at high frequencies than a 68μF capacitor would, because of its smaller lead inductance. All the capacitors except the 68μF Tantalum capacitors are surface-mount. The lengths of the bias and supply lines on the PCB shown in Fig.84 were tried to be minimized to reduce the amount of interference they could pick and also the filtering components were placed as close to the chip as possible. Potentiometers were used on the PCBs in order to be able to adjust the voltages and the currents. The outputs of the VCOs are taken differentially through SMA connectors from the top and bottom of the PCBs. The other SMA connectors at the right and left sides of the PCBs are for the inputs and outputs of some other building blocks designed by a colleague. The outputs of the VCOs are connected to the SMA connectors through series C parallel L impedance matching circuits. It is recommended that the paths on the PCB carrying the VCO output signals be designed to have the required width for 50 Ω characteristic impedance wherever possible.

The graph shown in Fig.86 depicts the distribution of the measured oscillation frequency of the 28 ICs out of the total 30 received from the foundry. The two ICs not included are defective in terms of the control voltages they start to oscillate at. The distribution of the single-ended fundamental output power of the 28 ICs is given in Fig.87. A rather large spread at the output power is observed in the chart. Note that no amplitude control loop was embedded into the design. In order for this circuit to be reliably used in an industrial application, an amplitude control loop would have to be embedded so that the output amplitude is kept at a well defined level at the different chips despite the mismatches and the process spread involved.

The frequency spectrum of the NMOS VCO with HP 0.5μm is given in Fig.88. A frequency span of 6 MHz was used in the FSEB 30 Spectrum Analyzer, marked as "4" in Fig.80. The fundamental output power is -29.83dBm. Fig.89 on the other hand depicts the fundamental and second and third harmonics of the same VCO. The second and third harmonic power levels are -85.7dBm and -87.7dBm, respectively. A difference of more than 55 dB is observed between the fundamental and the second and third harmonics.

The frequency tuning characteristic of the NMOS VCO with HP 0.5μm is shown in Fig.90. The operating point voltage at the gate terminal of the bulk-tuned PMOS

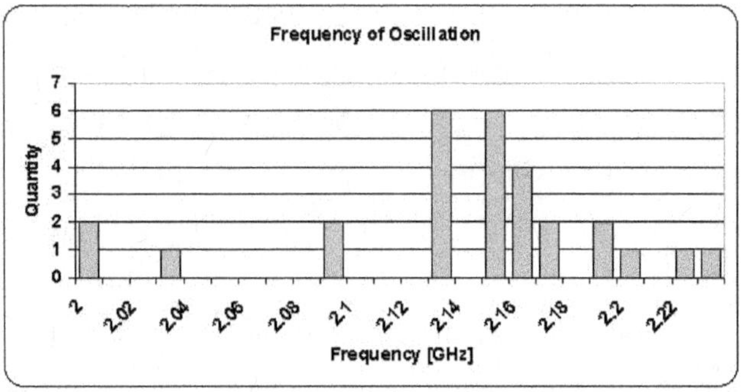

Fig. 86. Distribution of the oscillation frequency among the fabricated NMOS VCOs for $V_{bulk} = 2.5V$ with HP 0.5μm CMOS technology.

capacitor is at the supply voltage of 2.5V. Recall that the drain and source terminals of the PMOS used as the varactor are grounded. The variation of the fundamental output power over the tuning range is demonstrated in Fig.91.

It is seen in the plot that as the control voltage increases the capacitance of the varactor increases thus reducing the frequency of oscillation. This requires that the x axis be the voltage across the gate and the bulk of the PMOS. As the voltage across the gate and the bulk of the PMOS increases the channel under the gate goes from depletion to accumulation therefore the capacitance seen at the gate increases causing the frequency of oscillation to drop. The drop at the measured fundamental output power observed in Fig.91 shows that the loop gain of the VCO is decreased as the frequency is increased.

The VCO designed with HP 0.5μm CMOS draws 9mA from a 2.5V supply when oscillating at 2.14Gz. Fig.92 shows the measured phase noise of the VCO with the control voltage set at 2.5V. The low quality factor of the spiral inductors in the technology used (around 2.5 at 2.2GHz) limited the phase noise to a relatively high

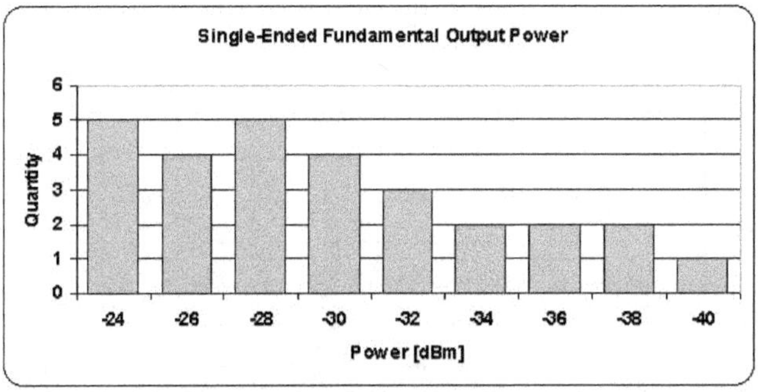

Fig. 87. Distribution of the single-ended fundamental power among the fabricated NMOS VCOs with HP 0.5μm CMOS technology.

value of -124dBc/Hz at 3 MHz offset from the carrier. The measured value is within 2dBs of the SpectreRF simulated value shown in Fig.18.

The phase noise of the VCO varies over the frequency tuning range due to the nonlinear capacitance-voltage transfer characteristic of the varactor and the variation of the quality factor of the varactor over its capacitance range. The phase noise of the VCO measured with the control voltage set at 1.3V is shown in Fig.93. A substantial degradation is observed at the phase noise at large offsets from the carrier with the phase noise at 3 MHz offset measured as -117dBc/Hz, which is 7dB higher than the measured phase noise at $V_{control} = 2.5V$. The considerable decrease at the fundamental output power with the control voltage of 1.3V seen in Fig.91 explains the degradation in the phase noise, as the phase noise is inversely proportional to the oscillation amplitude.

The die micrograph of the VCO fabricated with TSMC 0.35μm CMOS technology is shown in Fig.94. It occupies a silicon area of 0.26mm X 0.4mm. The circuit is part of the RF filter the measurements of which are presented in the last subsection. The

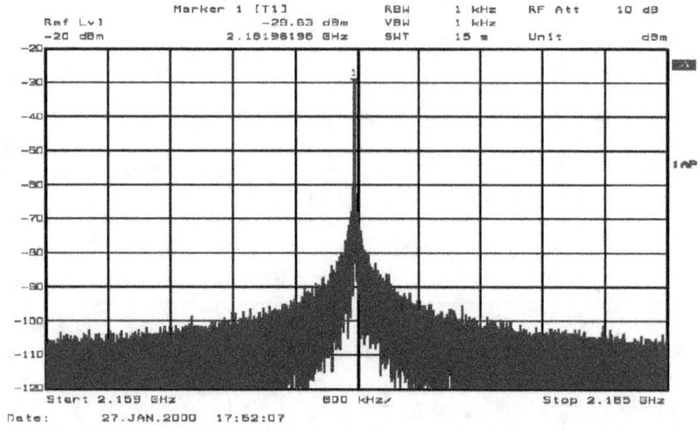

Fig. 88. Frequency spectrum of the HP 0.5μm NMOS VCO.

PCBs designed in a similar way as those designed for the LC VCO characterizations mentioned above are also explained in the last subsection. The frequency spectrum of the CMOS VCO with TSMC 0.35μm is plotted in Fig.95. A frequency span of 10 MHz was used to obtain this spectrum with the FSEB 30 Analyzer. A ZEL 1724LN Low Noise Amplifier from Mini Circuits was used to amplify the output power of the VCO to around 1 dBm for a more reliable phase noise measurement. The frequency of oscillation is 2.16GHz with the VCO core drawing 3.8mA from a 1.3V supply. In another frequency spectrum measurement, the frequency span was increased to allow for the inclusion of both the second and third harmonics, as shown in Fig.96. The second harmonic level was measured to be around 50 dB lower than that of the fundamental, whereas the power level at the third harmonic seemed to be below the noise floor. The tuning range of the VCO shown in Fig.97 is 18%, when the VCO core is drawing 6.8mA from a 1.8V supply. The measured tuning range is reduced

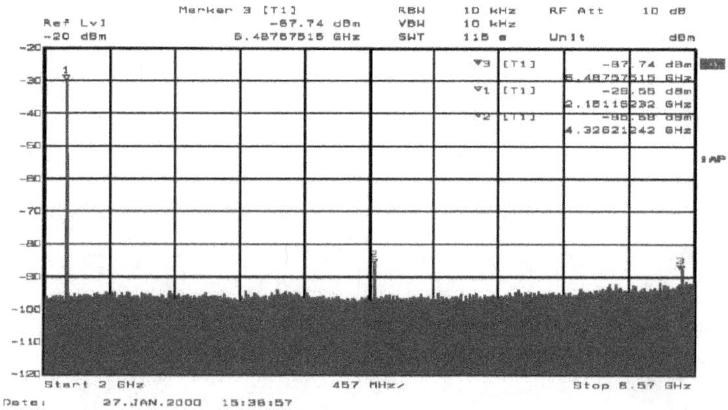

Fig. 89. Frequency spectrum of the harmonic contents of the HP 0.5μm NMOS VCO.

to around 4.8% with 1.3V supply. The limitation in the measured tuning range with the reduced supply voltage is observed to be due to the decreased tail current of the negative-conductance generating active devices because of the insufficient voltage headroom of the current mirror transistor. On-wafer S-parameter measurements of the PMOS inversion-mode varactor (configuration shown in Fig.49) used in the VCO core yielded a C_{max}/C_{min} ratio of around 2.3 at 2GHz. The phase noise of the VCO measured with the FSEB 30 Spectrum Analyzer and its dedicated Phase Noise Measurement Software called FSEK-04 (installed in the hard disk of the computer marked as "5" in Fig.80) is given in Fig.98. As the VCO was a part of the filter and no precautions were taken in the design phase to obtain low noise, the measured phase noise performance is very poor with a value of around -112dBc/Hz at 3MHz offset from the carrier of 2.15GHz.

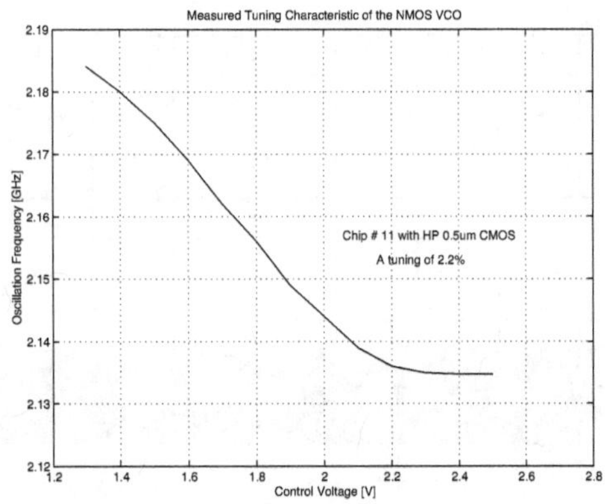

Fig. 90. Frequency tuning characteristic of the HP 0.5μm NMOS VCO.

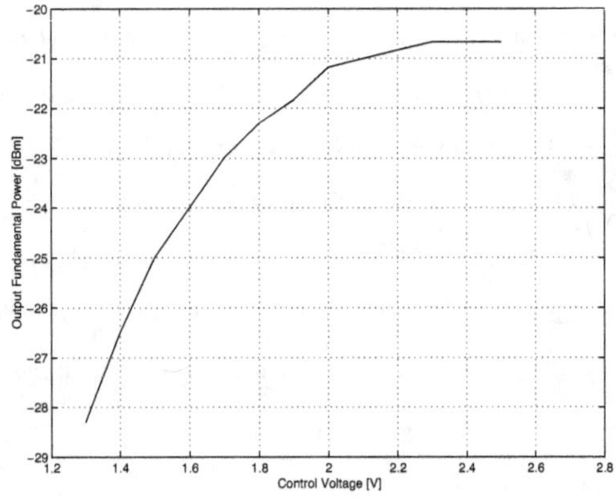

Fig. 91. Fundamental output power over the frequency tuning of the HP 0.5μm NMOS VCO.

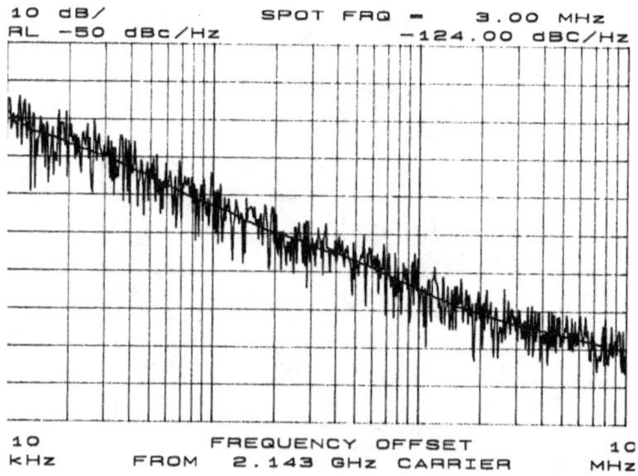

Fig. 92. Phase noise of the VCO with HP 0.5μm CMOS at $f_{osc} = 2.14GHz$.

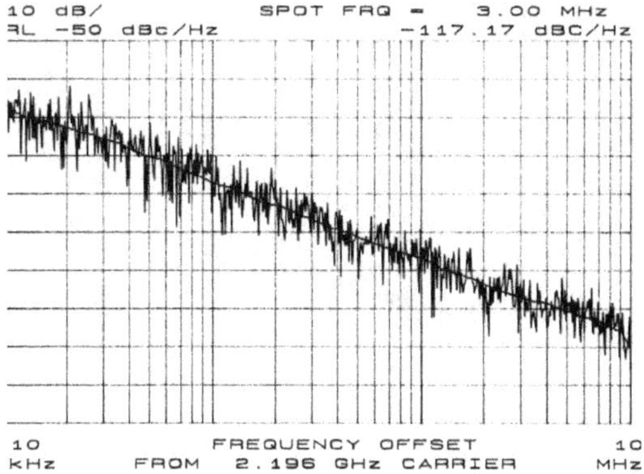

Fig. 93. Phase noise of the VCO with HP 0.5μm CMOS at $f_{osc} = 2.2GHz$.

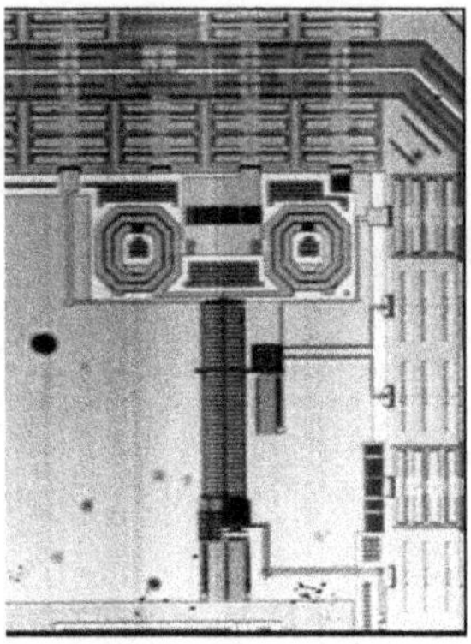

Fig. 94. Die micrograph (0.26mm X 0.4mm) of the TSMC 0.35μm CMOS VCO.

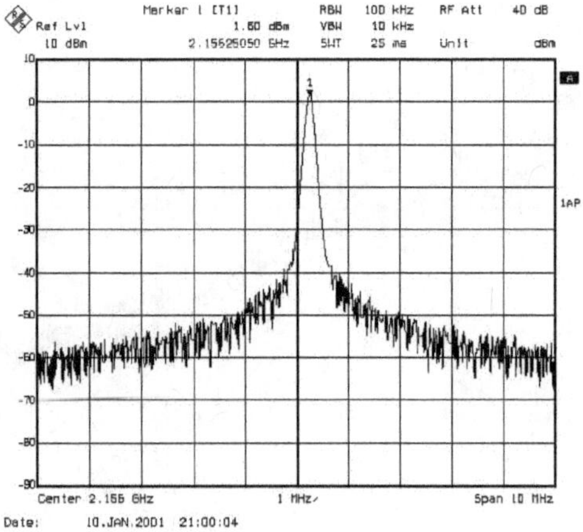

Fig. 95. Frequency spectrum of the TSMC 0.35μm CMOS VCO with Vdd=1.3V.

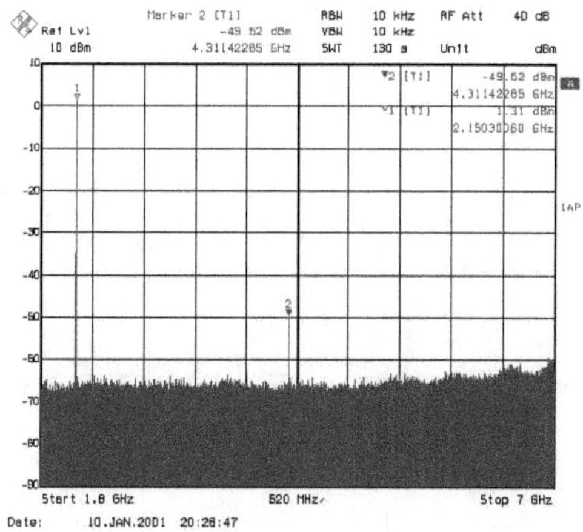

Fig. 96. Harmonic contents at the output of the TSMC 0.35μm CMOS VCO with Vdd=1.3V.

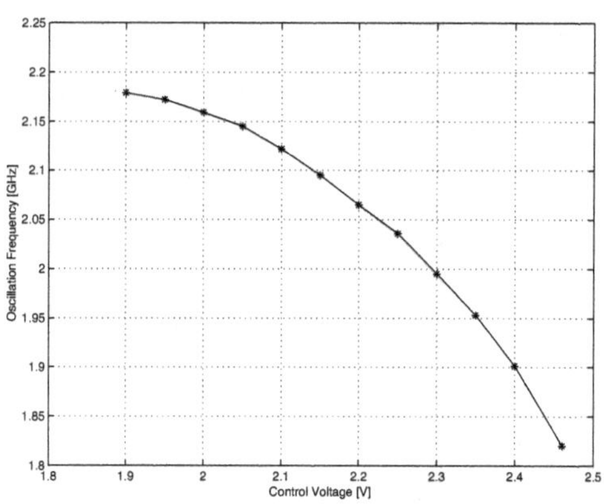

Fig. 97. Measured tuning range of the TSMC 0.35μm CMOS VCO with $V_{dd} = 1.8V$.

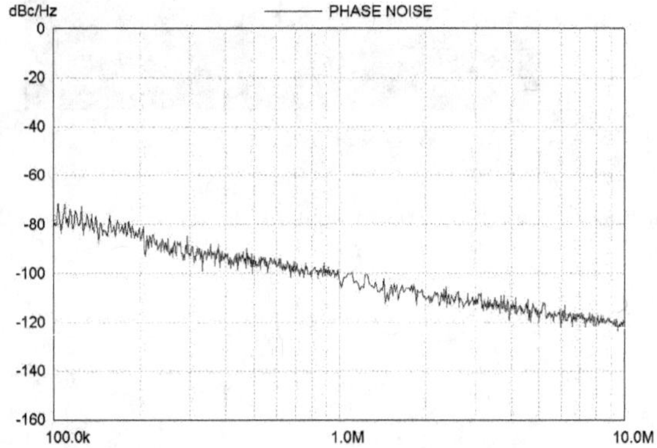

Fig. 98. Phase noise of the VCO with TSMC 0.35μm CMOS at $f_{osc} = 2.15GHz$.

2. A Symmetric Linearized CMOS LC VCO with Bulk-Tuned PMOS Capacitors

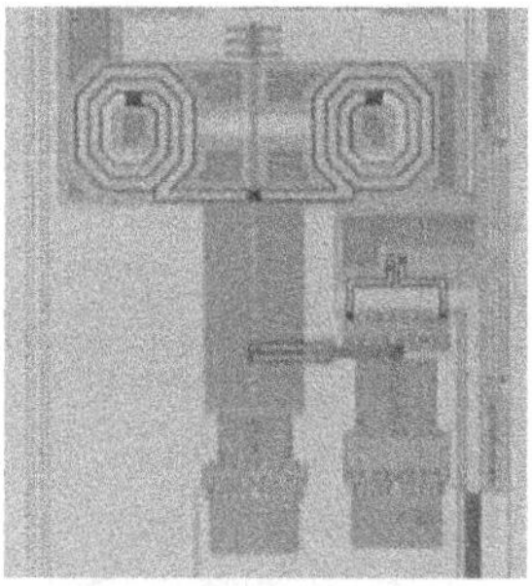

Fig. 99. Die micrograph (0.3mm X 0.4mm) of the CMOS voltage-controlled oscillator including the output buffer.

The CMOS VCO with the bulk-tuned PMOS capacitors and complementary cross-coupled differential pairs for negative conductance generation was fabricated with HP 0.5μm CMOS technology through MOSIS. The graph in Fig.100 depicts the distribution of the measured oscillation frequency of all the 30 ICs received from the foundry. The highest and lowest frequencies of oscillation are within $\pm$2% of the mid point, showing good consistency. The single-ended fundamental output power measured with the Spectrum Analyzer also demonstrates a reasonable consistency as shown in Fig.101.

The source degeneration resistors used in the NMOS cross-coupled pair for linearization are 50 Ω poly resistors, whereas no source degeneration resistors are employed in the PMOS cross-coupled pair in the circuit implemented. The technology is a standard digital technology with three metal and one poly layers. The chip photomicrograph is shown in Fig.99. The measurements of the VCO were performed on packaged chips. A simple differential pair with on chip poly resistor loads was used as an output buffer to drive the output pads bonded to a TQFP44 package. The output

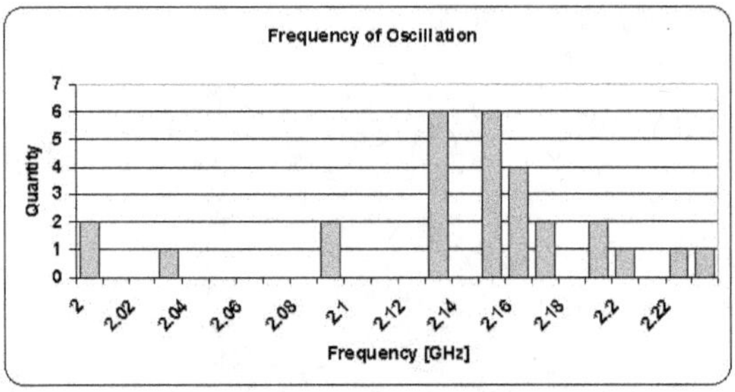

Fig. 100. Distribution of the oscillation frequency among the fabricated CMOS VCOs for $V_{bulk} = 2.5V$ with HP $0.5\mu m$ CMOS technology.

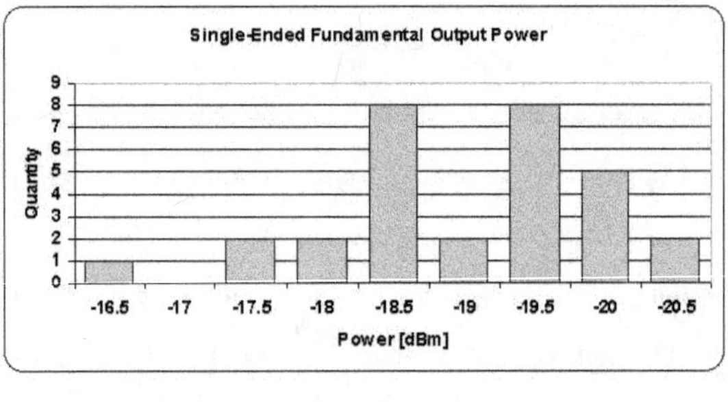

Fig. 101. Distribution of the single-ended fundamental power among the fabricated CMOS VCOs with HP $0.5\mu m$ CMOS technology.

buffer loads the VCO with a capacitance of around 160fF at each branch and draws 4mA from a separate 3V supply.

The frequency spectrum of the CMOS VCO measured with Rohde-Schwarz FSEB 30 spectrum analyzer was plotted in Fig.102. The frequency of oscillation is 2.2GHz. The harmonic contents of both the single-ended and differential outputs of the VCO are demonstrated in Figs.103 and 104, respectively. A ZAPDJ-2 Power Combiner from Mini Circuits was used to convert the two single-ended outputs of the VCO

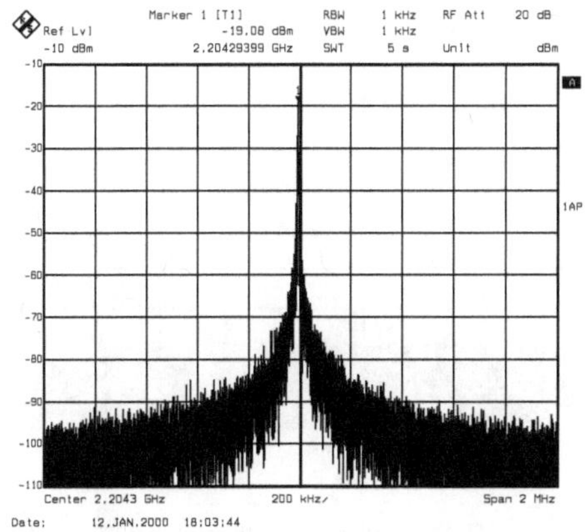

Fig. 102. Measured frequency spectrum of the CMOS VCO.

into a differential output to be applied to the analyzer. Ideally, the fundamental power should increase by 3dB and the second harmonic level should vanish once the single-ended to differential conversion is made. Due to the mismatches at the outputs and the nonidealities at the power combiner, an increase of around 1.5dB at the fundamental power level and a decrease of around 8.5dB at the second harmonic power level are measured with the single-ended to differential conversion, as shown in Figs.103 and 104. The phase noise of the VCO was characterized using the built-in phase noise measurement utility of an HP8563E spectrum analyzer. The low Q inductors limit the phase noise of the VCO drawing 4mA from a 3V supply to -126dBc/Hz at 3 MHz offset from a carrier of 2.19GHz, with the bulk voltage of the bulk-tuned PMOS varactors set at 3V, as shown in Fig.105. A comparison between the simulated and measured phase noise depicted in Fig.106 shows good consistency, with a maximum deviation of less than 4% within the offset frequency range of 1kHz

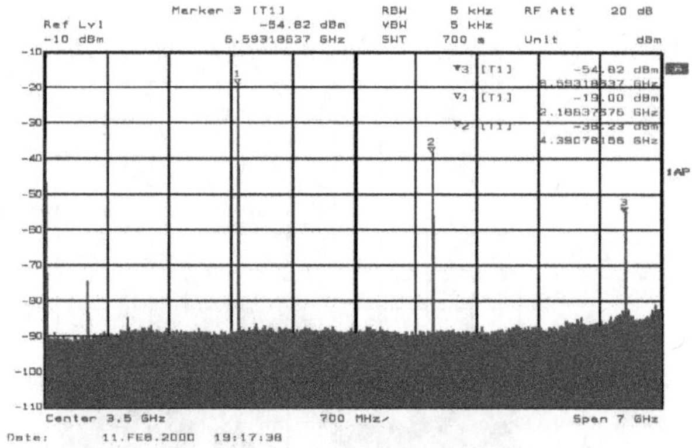

Fig. 103. Measured harmonic content of the single-ended output of the CMOS VCO.

to 3MHz from the carrier. The measured and simulated $1/f^3$ corner frequencies are at around 55kHz and 60kHz, respectively. The calculated phase noise of the VCO was also plotted on the same figure. The better matching between the calculations and the measurements at the $1/f^3$ spectrum of the phase noise indicates that the accuracy in the determination of the dc values of the ISFs of the components is somehow superior to the accuracy in the determination of the mean-square values of the ISFs of the components, through the charge injection simulations performed.

A figure-of-merit, [111], for an LC VCO design that provides a performance measure normalized in power consumption and tank quality factor is given in (5.1)

$$FOM = \left(\frac{f_o}{\Delta f}\right)^2 \cdot \frac{1}{\mathcal{L}(\Delta f) \cdot P_D \cdot Q^2} \tag{5.1}$$

where f_o is the frequency of oscillation, Δf is the frequency offset from the carrier, P_D is the power consumption of the VCO in mW, Q is the tank quality factor and $\mathcal{L}$ is the phase noise. The quality factor of an integrated tank has a strong dependence on the available technology, therefore it is justifiable to normalize the figure-of-merit

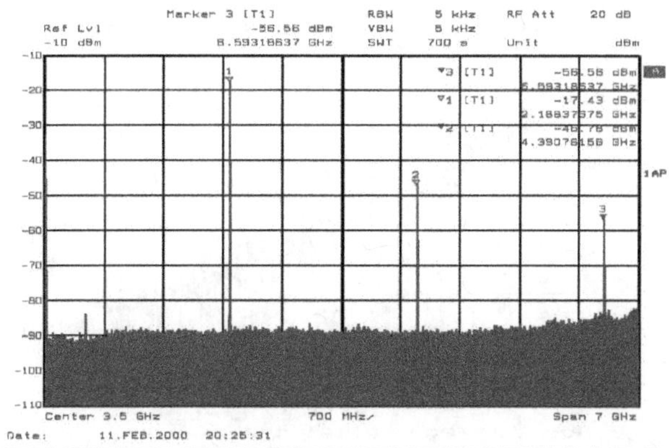

Fig. 104. Measured harmonic content of the differential output of the CMOS VCO.

of a VCO with respect to the quality factor of its tank for a fair comparison. A comparison with some of the CMOS and Bipolar VCOs in the literature designed with Si or SiGe technologies is provided in Table VIII.

Note that the VCO presented in this paper shows a reasonable performance based on the normalized figure-of-merit given above, compared to the other state-of-the-art VCOs in the existing literature. The tuning range of the VCO as shown in Fig.107 is 3.5% due to the limitation of the well resistance in the bulk-tuned varactors. The output power of the fundamental also varies through the frequency tuning range as shown in Fig.108. The phase noise changes throughout the tuning range due to the fact that the quality factor of the bulk-tuned PMOS capacitors which provide the frequency tuning, varies and modifies the overall quality factor of the tank, as shown in Fig.109. The variation of the phase noise over the frequency tuning range is consistent with the variation of the fundamental output power illustrated in Fig.108. The comparison between the simulated and measured phase noise of the VCO for two

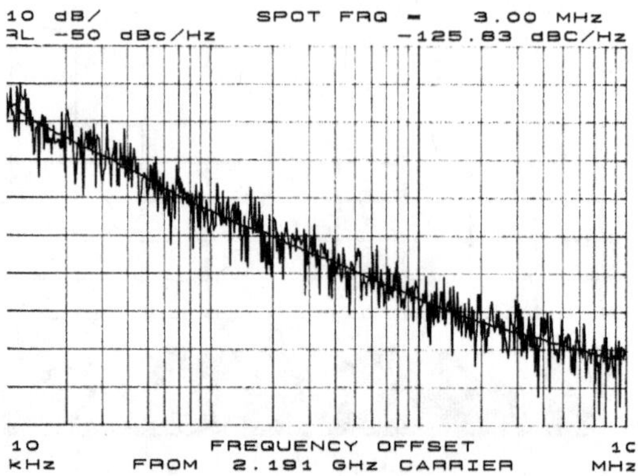

Fig. 105. Measured phase noise of the CMOS VCO at $f_{osc} = 2.19GHz$.

different varactor control voltage settings are given in Fig.110 and Fig.111.

The match between the simulated and measured phase noise for three different bulk voltage (control voltage) settings at frequency offsets larger than 100kHz is satisfactory. This clearly shows that the effect of the low Q LC tank is well captured in the simulations. The discrepancy between the simulated and measured phase noise at offset frequencies lower than 100kHz for $V_{bulk} = 1.5V$ shown in Fig.110 is anticipated to be due to inadequate modeling of the capacitance-voltage transfer characteristic of the PMOS varactor in the simulations. The higher simulated phase noise at the $1/f^3$ region suggests that the nonlinearities in the VCO governing the $1/f$ noise upconversion are not modeled accurately. As the match between the simulated and measured phase noise at other bulk voltage settings are satisfactory, it is concluded that the discrepancy at low offsets in Fig.110 originates from the inadequate modeling of the PMOS bulk-tuned varactor's nonlinear capacitance-voltage characteristic around $V_{bulk} = 1.5V$.

Note that in this IC, the source-degeneration resistors were used only in the

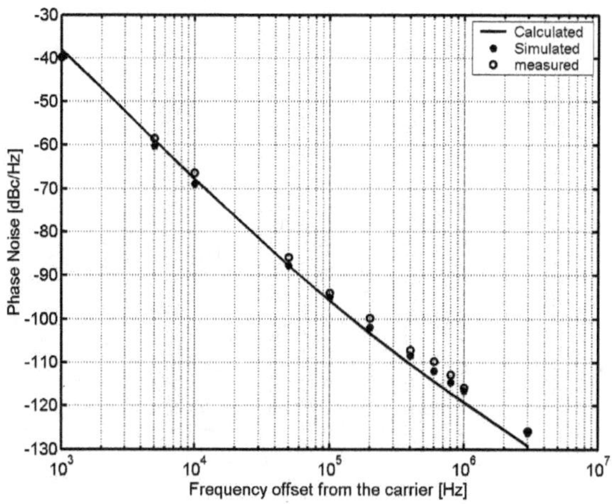

Fig. 106. Measured vs. simulated and calculated (with Eqs. (2.17) and (2.21)) phase noise of the CMOS VCO.

NMOS cross-coupled transistors. In a more recently fabricated chip, all three configurations mentioned in Chapter II Section C, were housed in order to provide experimental verification of the design trade-offs pointed out from the simulations. The source-degeneration resistors used in the VCOs are 50Ω resistors built with the poly layer of the technology, which is a standard digital technology with three metal and one poly layers. The photomicrograph of the chip that includes the three VCOs is shown in Fig.112. The measurements of the VCOs were performed on packaged chips. Simple differential pairs with on chip poly resistor loads were used as output buffers to drive the output pads bonded to a TQFP44 package in all the VCOs. The output buffer loads the VCO with a capacitance of around 160fF at each branch and draws 4mA from a separate 3V supply.

The phase noise measurements of the three VCOs included in the second chip

Table VIII. Calculated normalized Figure-of-Merit (FOM) for some CMOS and Bipolar VCOs in the literature.

Reference	Q	f_o [GHz]	P_D [mW]	$\mathcal{L}$ [dBc/Hz]	FOM [dB]
[36]	9	1.55	21.6	-102@100kHz	153
[112]	4	1.96	32.4	-136@4.7MHz	161
[57]	8	1.9	21.6	-123@600kHz	162
[113]	16	0.8	4.3	-106@100kHz	154
[111]	8	2.56	14	-104@100kHz	163
[53]	6.5	1.91	10	-121@600kHz	165
[114]	10	1	9.1	-152@3MHz	173
[115]	8.5	50	13	-100@1MHz	164
[40]	11	2	34.2	-126@600kHz	160
[55]	7.5	1.8	6	-121@600kHz	165
[116]	9	1.96	19.5	-150@3MHz	174
This work	2.5	2.19	12	-125.83@3MHz	164

were performed using an RDL NTS-1000B phase noise measurement system. Two different chips were measured and the results plotted in the following figures are the averaged noise performances of the two chips. Each chip was measured 10 times consecutively for all the VCOs and since two different chips were measured, the total amount of measurements averaged for each VCO was 20. The comparisons plotted in Figs. 113 and 114 demonstrate the impact of the source-degeneration resistors on the phase noise at close-in offsets and at the large offsets from the carriers, respectively. The improvement at the close-in offsets as the resistors are added is obvious. On the other hand, as the offset from the carrier gets larger, the degradation in phase

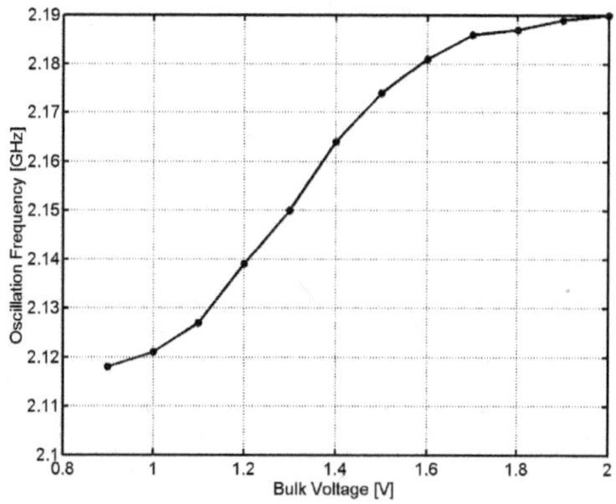

Fig. 107. Measured tuning range of the CMOS bulk-tuned VCO.

noise is observed as the resistors are added due to the additional noise of the resistors and because of the reduction of the oscillation amplitude for the same bias current used in the designs. Note that at the largest offset in Fig. 114, the phase noise of the VCO without any resistors is still slightly larger than that of the VCO with resistor at the NMOS pair only. However, at even larger offsets, the phase noise slopes show that the VCO without any resistor would have less noise than the other two as the simulations, depicted in Fig.33 up to offsets of 10MHz from the carrier, suggested. The RDL NTS-1000B phase noise measurement system allows 1MHz as the largest offset from the carrier in measurements. It can be stated that the phase noise measurements agree well with the simulations regarding the trade-offs involved in the use of source-degeneration resistors.

The tuning ranges of the VCOs are shown in Fig.115, which demonstrates the strong reduction at the frequency tuning capability of the circuits as the source-degeneration resistors are added with the same bias current due to the reduction

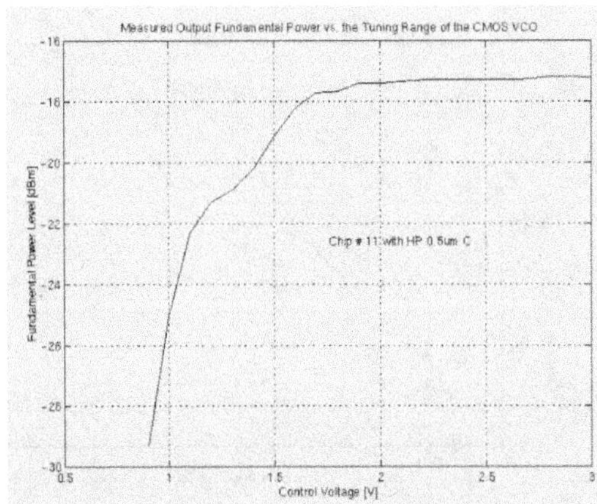

Fig. 108. Variation of the single-ended fundamental output power over the tuning range of the CMOS bulk-tuned VCO.

of the negative conductances generated in the cross-coupled pairs. The phase noise changes throughout the tuning range due to the fact that the quality factor of the bulk-tuned PMOS capacitors varies and modifies the overall quality factor of the tank and thus the oscillation amplitude, as shown in Fig.116 at the large offset of 1MHz from the frequency of oscillation. The plot depicts the phase noise of the VCO with source-degeneration resistors at both the NMOS and the PMOS cross-coupled pairs. To summarize, the symmetric source-degeneration resistors reduce the close-in phase noise of the CMOS VCO with sacrifices in the phase noise at large offsets from the carrier and in the frequency tuning range for a given bias current.

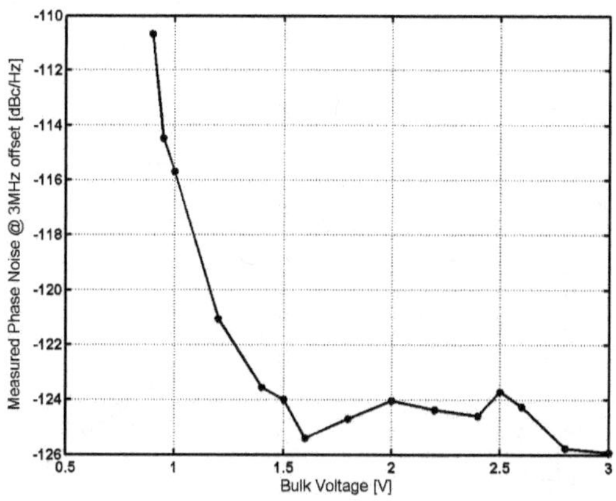

Fig. 109. Variation of the measured phase noise at 3MHz offset throughout the tuning range.

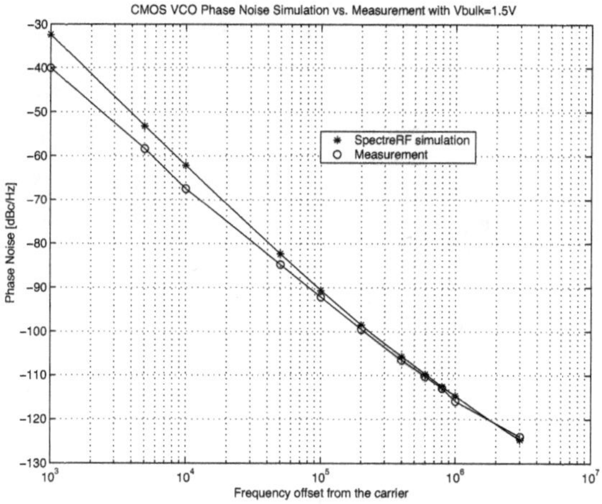

Fig. 110. Comparison between the simulated and measured phase noise with $V_{bulk} = 1.5V$.

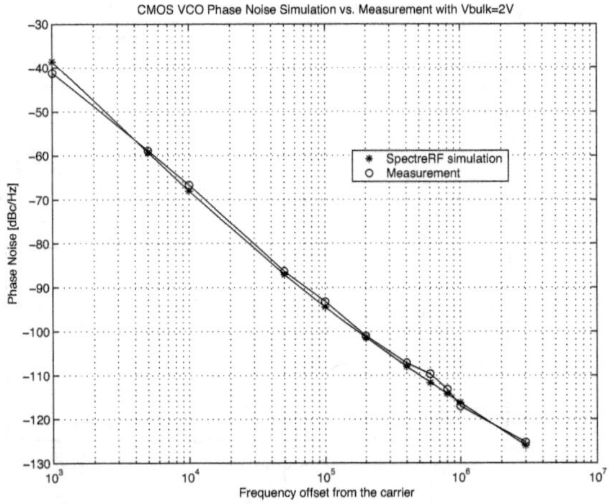

Fig. 111. Comparison between the simulated and measured phase noise with $V_{bulk} = 2V$.

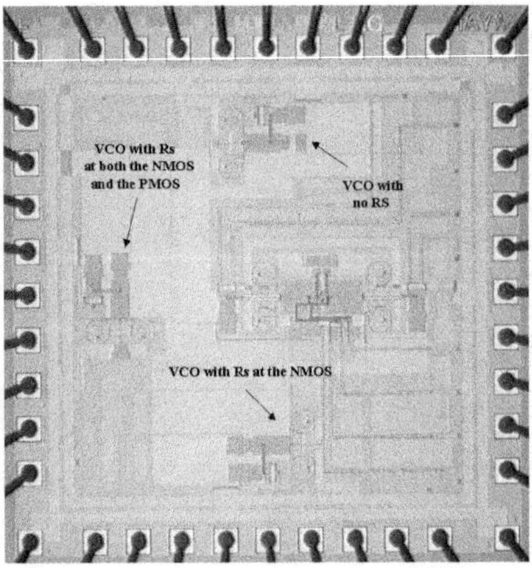

Fig. 112. Die micrograph of the chip including the three CMOS voltage-controlled oscillators.

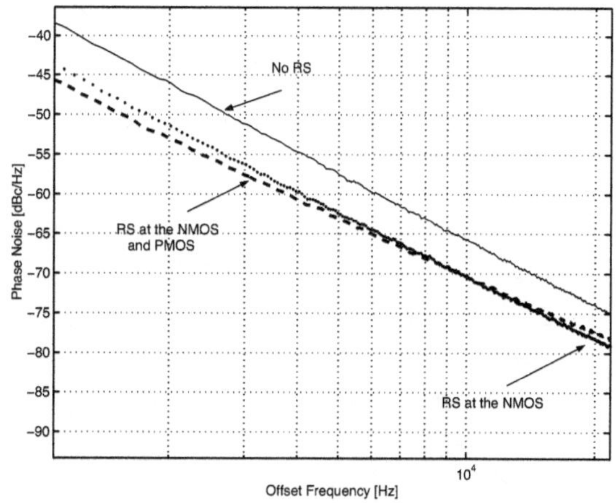

Fig. 113. Measured effect of the source degeneration resistors on the close-in phase noise of the CMOS VCOs.

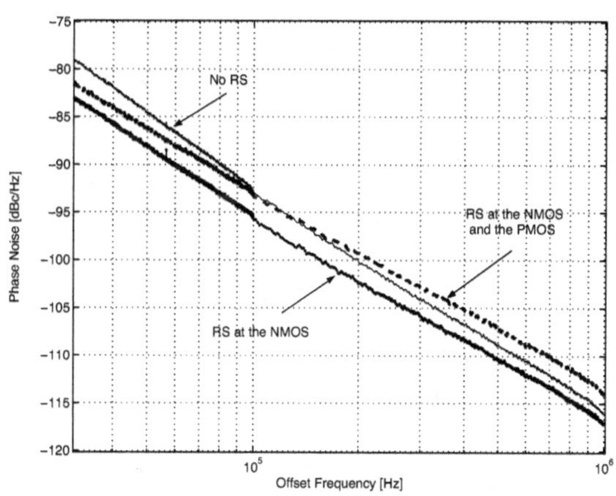

Fig. 114. Measured effect of the source-degeneration resistors on the phase noise at large offsets from the carrier of the CMOS VCOs.

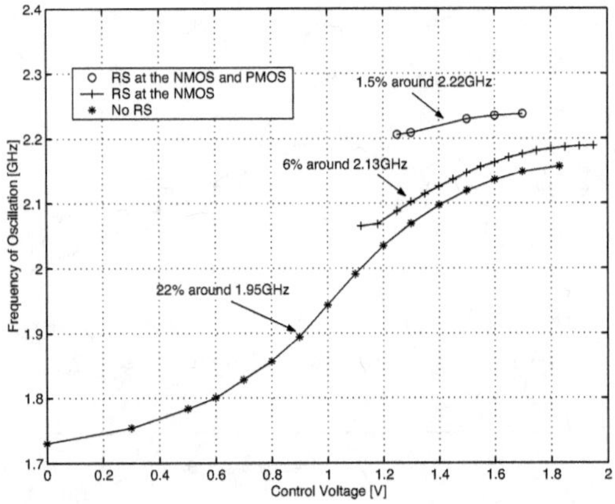

Fig. 115. Comparison of the measured tuning ranges of the VCOs.

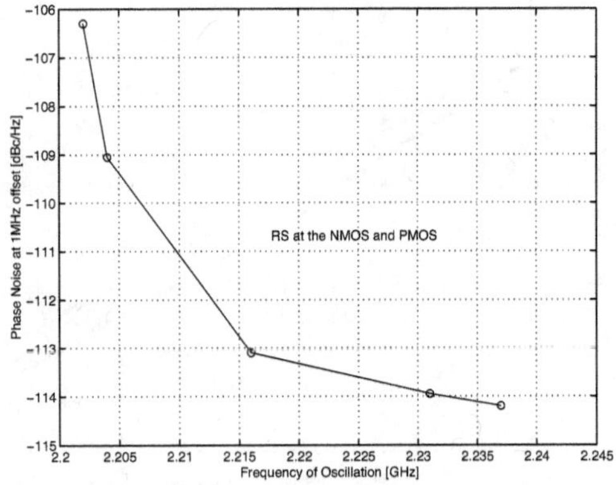

Fig. 116. Variation of the measured phase noise at 1MHz offset throughout the tuning range.

B. A Dual-Modulus Prescaler in 0.6 μm BiCMOS

A dual-modulus prescaler was designed and fabricated with a 0.6μm BiCMOS process of Texas Instruments. The technology provides NPN transistors with a peak f_T of 22 GHz. The die micrograph of the prescaler is shown in Fig.117. The prescaler including the bias network and the output buffer occupies around 375μm X 350μm of Silicon area excluding the bonding pads.

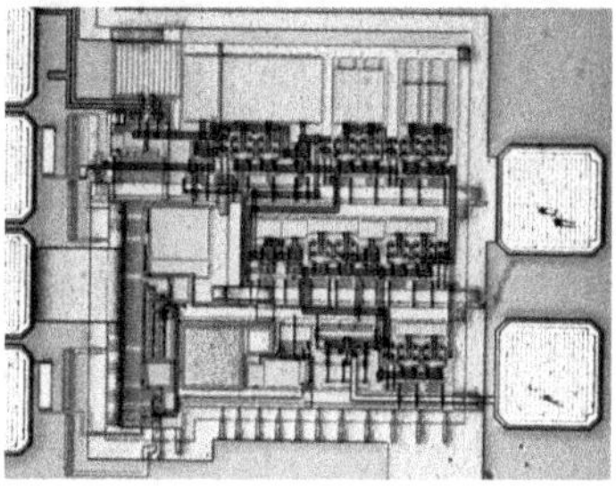

Fig. 117. Die micrograph (0.38mm X 0.35mm) of the prescaler including the bias network and the output buffer.

The PCB designed for testing the prescaler ICs is shown in Fig.118. The differential to single-ended conversion was performed on the PCB using an ETC1-1-13 BALUN from MACOM. The prescaler processes a single-ended RF input. The input buffer converts the single-ended input into differential signals to drive the divide by 4/5 finite state machine. The input and the output of the circuit are through the SMA connectors on the PCB. There are four DC bias and supply voltage inputs for the IC: The supply voltage input for the prescaler; the supply voltage input of the output buffer; the modulus control input which enables divide-by-32 mode with a CMOS Logic 0 level (0V) and divide-by-33 mode with a CMOS Logic 1 Level (2.5V), and finally the power enable input that either enables or disables the bandgap reference circuit that generates the bias voltages and currents of the prescaler.

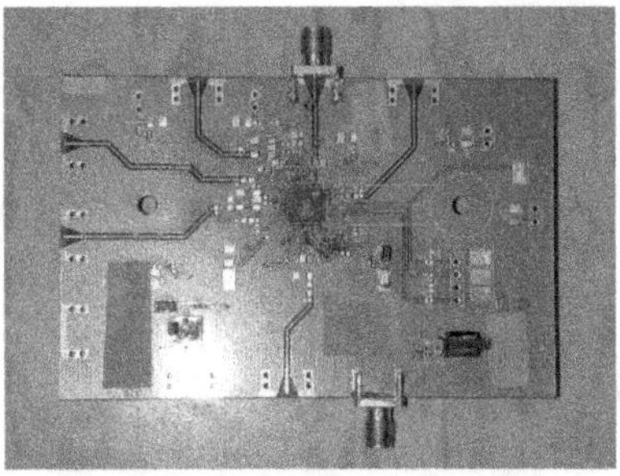

Fig. 118. PCB designed for the characterization of the prescalers.

The measurements were taken from parts packaged in Thin Quad Flat Packages (TQFP) with 44 pins. All the bias and supply voltage lines have surface mount bypass capacitors for noise filtering. A 50Ω resistor between the input and the ground before a DC blocking capacitor provides impedance matching for the RF input. A so called hand-down socket was used to test the functionality of all the ICs. The socket has a housing where the IC to be tested is placed in to be aligned with the soldering islands for the TQFP 44 package. Then there is a lid at the top part of the socket with an extension that pushes the package to contact the soldering islands by screwing. Out of the 15 packaged parts measured, 13 of them were found functional, giving a functionality yield of more than 85%. The maximum operating frequency and the power consumption of the functional parts are roughly within ±5% of the mean values as shown in the distribution given in Fig.119. One of the ICs that function properly was soldered on the PCB for a thorough characterization. The circuit draws 2.35mA from a 2.5V supply. The frequency spectrum of the prescaler is plotted in Fig.120.

The differential outputs of the prescaler were taken out of the chip through an emitter-coupled pair buffer with on-chip resistor loads. The buffer is self-biased with

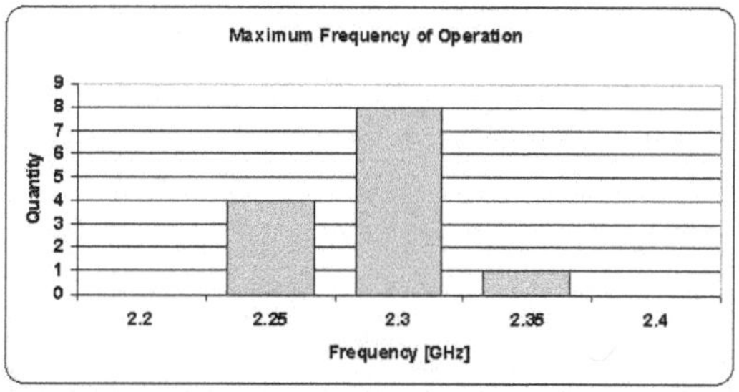

Fig. 119. Distribution of the maximum frequency of operation among the fabricated prescalers.

an on-chip bias resistor and it draws around 1.6mA from a separate power supply of 2.5V. The differential outputs are converted into a single-ended signal with a surface mount Balun on the test PCB for the measurement equipment. The transient output waveform of the prescaler for an input at 2.35GHz is shown in Fig.121 as monitored on an oscilloscope.

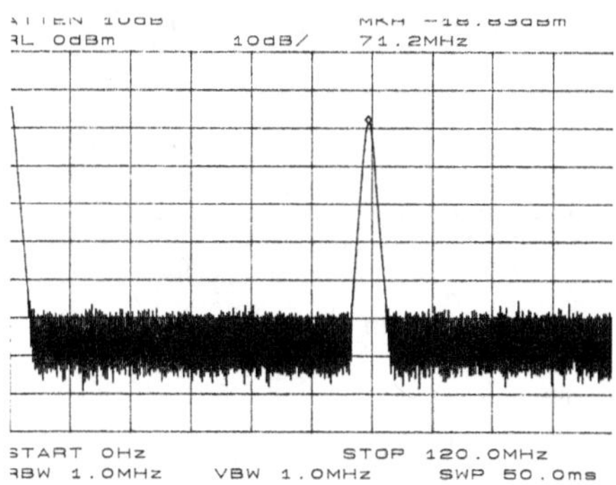

Fig. 120. Frequency spectrum of the prescaler with V_{cc}=2.5V.

The input sensitivity measurements of the prescaler in divide by 33 mode were

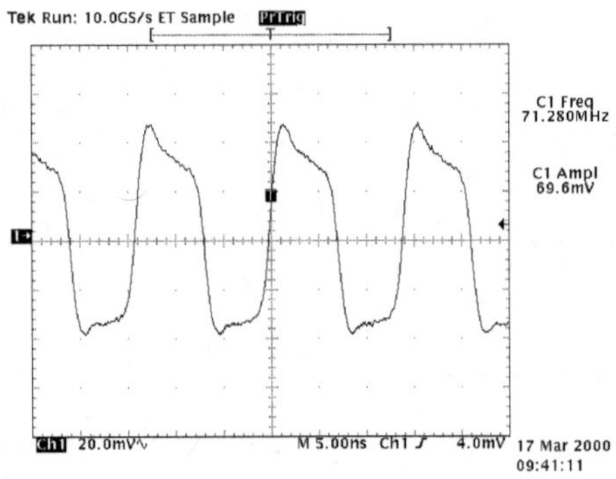

Fig. 121. Output of the prescaler with $f_{in} = 2.35$GHz.

done with V_{cc}=2.5V at room temperature. The proper division of the circuit was observed from a spectrum analyzer. The measurements show a wideband frequency range from 100 MHz to 2.4GHz as shown in Fig.122. The increase at the sensitivity at around 1.8GHz is due to the self-oscillation phenomena in the absence of an input signal, as commonly seen in the ECL-based dividers, [42]. The solid lines indicate the maximum and the minimum signal levels required for proper operation at the respective input frequencies. To determine the minimum supply voltage for the prescaler, the input sensitivity versus the supply voltage characteristic was also measured. As plotted in Fig.123, the circuit still operates with a supply voltage of 2.1V, but with quite a limited input sensitivity.

The phase noise of the prescaler was measured with an RDL NTS-1000B phase noise measurement system. Phase noise values of -107.6dBc/Hz and -120.1dBc/Hz were measured at 1kHz and 5kHz offsets, respectively, as shown in Fig.124. The values correspond to contributions of around -77dBc/Hz and -89.7dBc/Hz at 1kHz and 5kHz offsets to the phase noise of a PLL using this prescaler. In order to refer

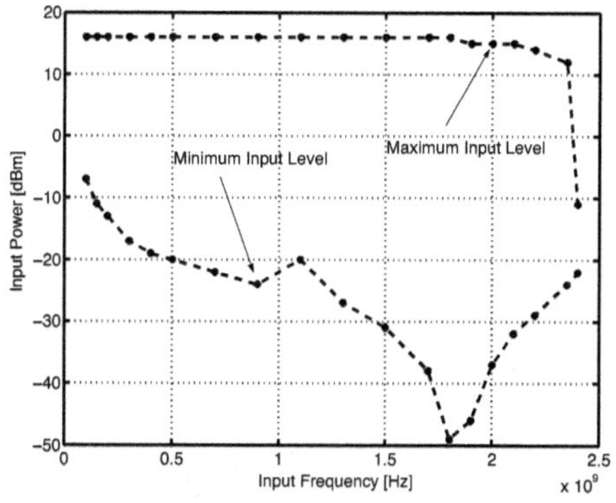

Fig. 122. Measured input sensitivity over the frequency.

the measured output phase noise of the prescaler to its input which is the output of the VCO and thus that of the PLL, the following formula is used:

$$\mathcal{L}_{in,ref} = \mathcal{L}_{out} + 10 \log N^2 \tag{5.2}$$

where $\mathcal{L}_{in,ref}$ and $\mathcal{L}_{out}$ denote the phase noise referred to the input of the prescaler and the output phase noise of the prescaler, respectively. N is the divide ratio set by the Modulus Control input of the IC. The noise at 1kHz offset is short of meeting the DCS specification, which requires less than -85dBc/Hz contribution from the prescaler. The circuit requires noise optimization to be done using the guidelines provided in Chapter III. In the same figure, the simulated phase noise is also plotted. It should be repeated that the main contributors to the phase noise at low offsets are found in the simulations to be the 1/f noise of the MOS current mirror transistors providing the bias to the divide by 8 extender asynchronous dividers. The diode-connected NMOS tail current mirror transistor biasing the asynchronous dividers contributes almost 62% of the total noise at 1kHz and 5kHz offsets. The bias currents of the building

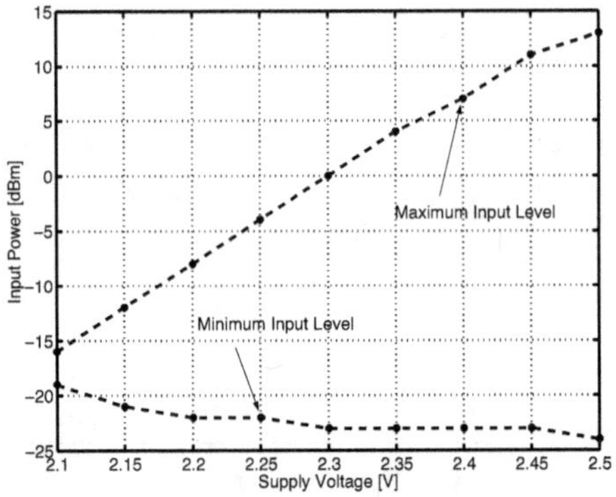

Fig. 123. Measured input sensitivity over the supply voltage.

blocks are distributed from the bandgap bias generator using PMOS current mirror transistors and the 1/f noise of the PMOS transistors providing the bias to the tail current mirror of the extender asynchronous dividers accounts for about 12% of the total phase noise at both 1kHz and 5kHz offsets. The match between the simulated and the measured phase noise is within 5dB for offsets ranging from 1kHz to 1MHz. Note that good consistency is observed between the simulated and measured data within the 1/f noise region. The phase noise of the prescaler was also measured with an HP8563E Spectrum Analyzer, as shown in Fig.125. The plots given in Figs.126-128 are the phase noise measurements of the prescaler averaged on multiple consecutive measurements taken with RDL NTS-1000B for three different test settings given in the figure captions. The comparison between three cases plotted in Fig.129 shows that the divide-by 32 and 33 modes yield very similar phase noise performance with the same input frequency of 2.35GHz, whereas in the case of smaller input frequency of 2GHz, as the output is at a lower frequency, a lower noise floor is measured.

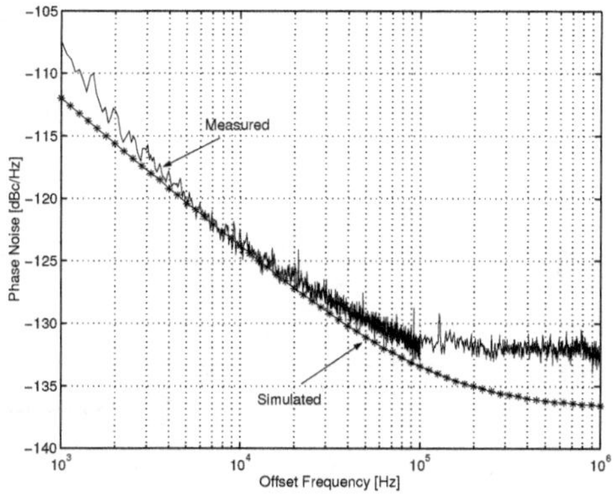

Fig. 124. Measured vs. simulated phase noise of the prescaler in divide-by-33 mode with $f_{in} = 2.35$GHz.

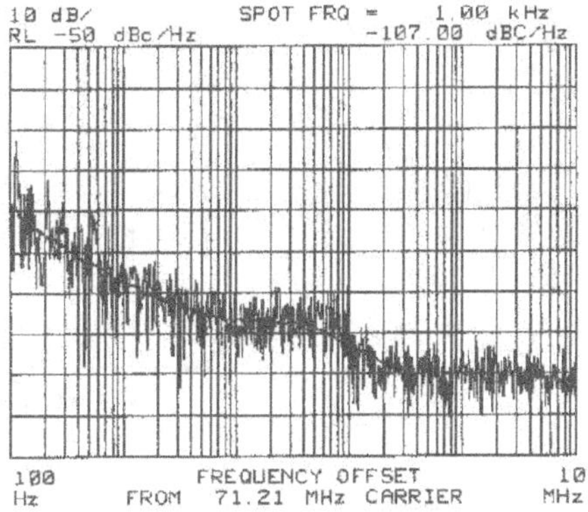

Fig. 125. Phase noise of the prescaler in divide-by-33 mode with $f_{in} = 2.35$GHz.

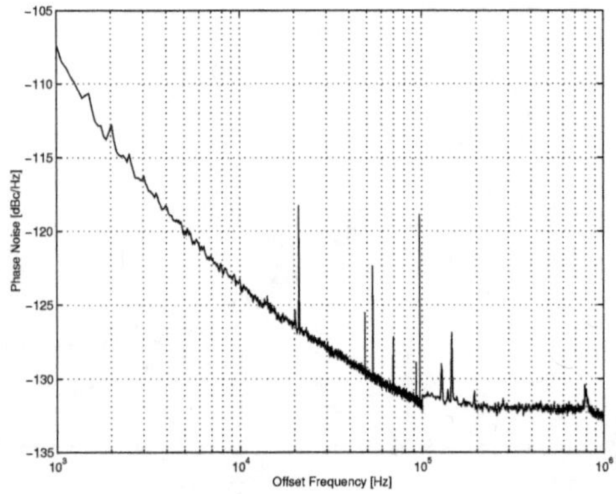

Fig. 126. Phase noise of the prescaler averaged on ten consecutive measurements in divide-by-33 mode with $f_{in} = 2.35$GHz.

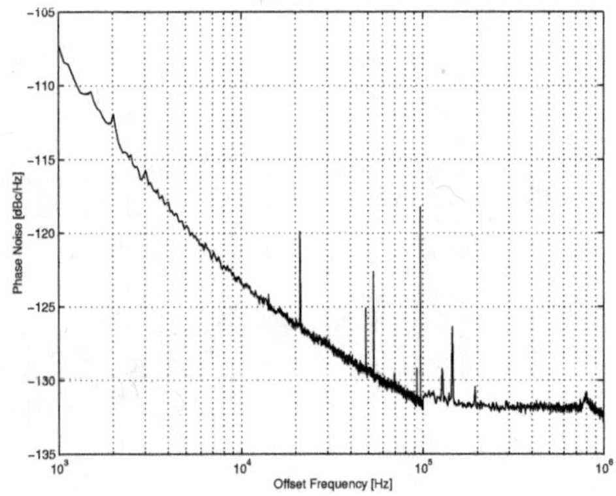

Fig. 127. Phase noise of the prescaler averaged on eight consecutive measurements in divide-by-32 mode with $f_{in} = 2.35$GHz.

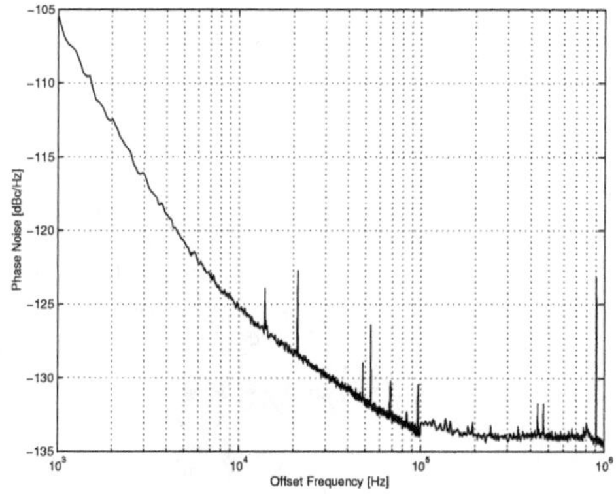

Fig. 128. Phase noise of the prescaler averaged on twelve consecutive measurements in divide-by-32 mode with $f_{in} = 2\text{GHz}$.

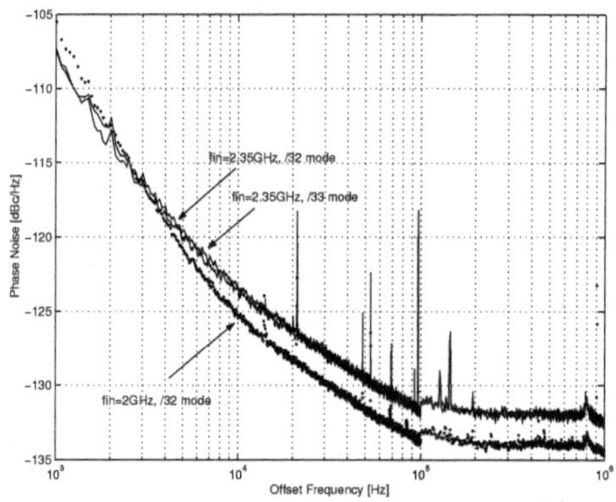

Fig. 129. Phase noise of the prescaler for three different test settings.

C. A Fully-Integrated LC Q-Enhancement RF Bandpass Filter in CMOS

The filter was fabricated with TSMC 0.35μm CMOS technology through MOSIS. The technology is a mainstream CMOS technology with four metal and two poly layers. The second poly layer was not used in the design, showing the compatibility with a standard Digital CMOS technology. The chip photomicrograph is shown in Fig.130. The filter occupies a silicon area of 0.4mm X 0.26mm including the output buffer.

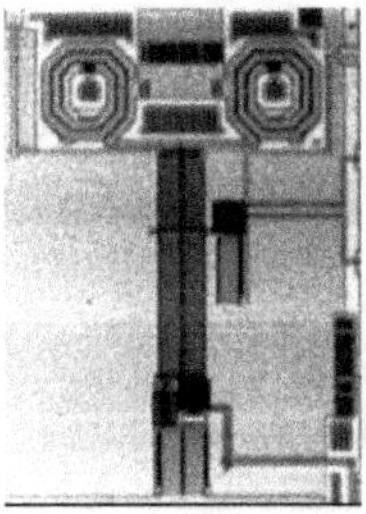

Fig. 130. Die micrograph (0.26mm X 0.4mm) of the CMOS bandpass filter including the output buffer.

The measurements of the filter were performed on chips enclosed in TQFP44 packages. A simple differential pair in open-drain configuration was used as an output buffer. The test setup used for characterizing the filter is shown in Fig.131. The open-drain output buffer is terminated on the test PCB with surface-mount resistors connected to a separate power supply than that of the filter core.

The bias voltage through the surface-mount resistors of 2.2K each at the input provide the gate bias to the differential pair transistors of the input g_m stage. The capacitors at the input and the output serve as DC blocking capacitors. Resistors not shown in the schematic are used as impedance matching components to provide relatively broadband matching. ETC1-1-13 baluns from MACOM provide the single-ended to differential and vice versa conversions.

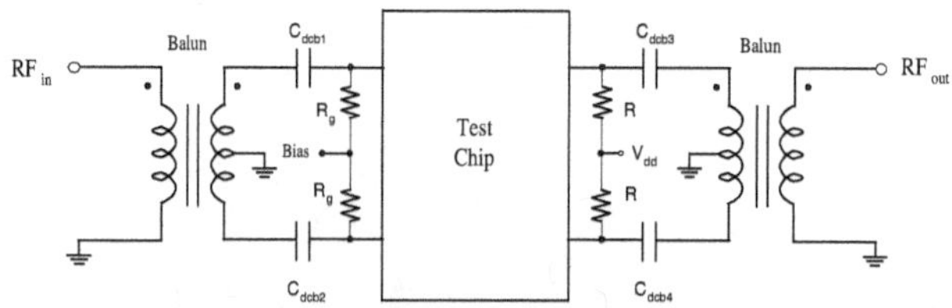

Fig. 131. Test setup for the filter measurements.

Fig. 132. PCB designed for the RF characterization of the filter.

Two PCBs were designed to properly characterize the RF filter ICs. Eleven ICs were characterized. Initially, the RF characterization of the filter was targeted to be performed on an IC soldered on the PCB designed for this purpose, shown in Fig.132. Another PCB designed with a 40 pin ZIF socket to roughly characterize all of the ICs received from the foundry is shown in Fig.133. The relatively large parasitics associated with the ZIF socket were believed to prevent a reliable RF characterization, but in contrary, some of the dynamic range figures presented in this section were measured with the ZIF socket. The same filtering RC sections depicted in Fig.85 for the LC VCO test PCBs were used on the bias, supply and control voltage lines on the PCB. It should be noted that LC matching circuits were avoided on the PCB of the RF filter in order not to interfere with the filter's frequency response. The outputs and the inputs of the filter and the standalone output buffer were taken through SMA female connectors from the PCB.

Fig. 133. PCB designed for the functionality test of the filter.

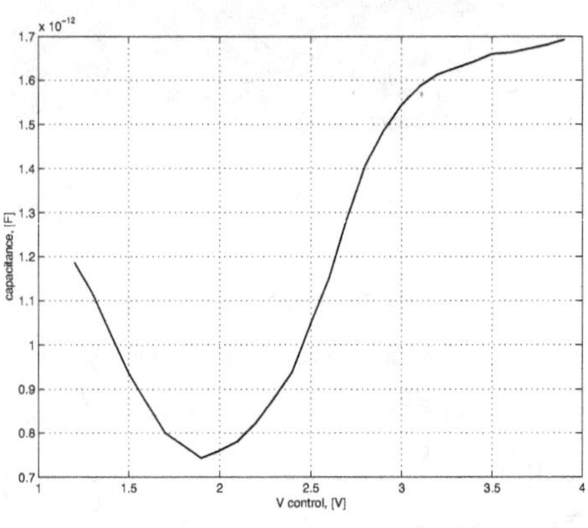

Fig. 134. Measured transfer characteristic of the PMOS varactor at 2GHz for V_{dd}=1.8V.

A stand-alone PMOS capacitor as those of the filter varactors was characterized with S-parameter measurements using Cascade Microprobes and an HP 8719ES Network Analyzer in the test setup shown in Fig.81. A capacitance ratio of 1:2.3 was measured at 2GHz, as shown in Fig.134. Note that the x-axis is the control voltage applied to the node where the drain, source and the bulk of the transistor are connected, while the gate voltage is kept constant at 1.8V, which is approximately the dc operating point of the output of the RF filter with V_{dd}=1.8V. The capacitance decreases as the channel approaches the deep depletion, making a minimum and then the inversion takes place gradually increasing the capacitance finally converging to the value of the gate oxide capacitance.

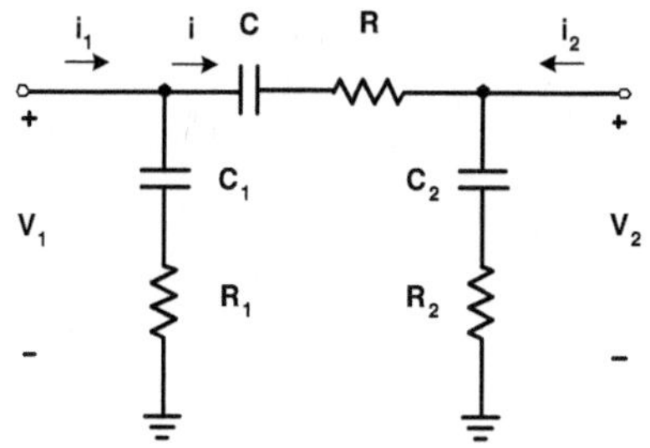

Fig. 135. Equivalent circuit of the PMOS varactor used in the capacitance extraction.

The S parameters of the PMOS varactor measured with the Network Analyzer were converted into y parameters and the capacitance value for each corresponding voltage across the PMOS biased using Bias Tees from Mini Circuits was extracted in accordance with the simplified equivalent circuit shown in Fig.135. The resistor R in series with the capacitor C that represents the capacitance of the varactor models the loss. The series combination of the capacitors and resistors at each side of the equivalent circuit model the capacitance between the associated terminal of the varactor and the substrate and the substrate resistances, respectively. The calculation of the component values in the equivalent circuit of Fig.135 using the y parameters is presented below. The admittance matrix (y parameters) is widely used to represent a two-port network, [110]. It relates the currents flowing into the ports to the port voltages in accordance with the following formula

$$\begin{bmatrix} i_1 \\ i_2 \end{bmatrix} = \begin{bmatrix} y_{11} & y_{12} \\ y_{21} & y_{22} \end{bmatrix} \begin{bmatrix} v_1 \\ v_2 \end{bmatrix}$$

$$\Rightarrow i_1 = y_{11}v_1 + y_{12}v_2 \quad (5.3)$$

$$i_2 = y_{21}v_1 + y_{22}v_2. \quad (5.4)$$

The node equations using the branch currents and the node voltages can be written from the equivalent circuit in Fig.135 as

$$v_1 - v_2 = i\left(\frac{1}{j\omega C} + R\right) \quad (5.5)$$

$$v_1 = (i_1 - i)\left(\frac{1}{j\omega C_1} + R_1\right) \quad (5.6)$$

$$v_2 = (i_2 + i)\left(\frac{1}{j\omega C_2} + R_2\right). \quad (5.7)$$

Solving (5.5) for i, substituting it into (5.6) and rearranging the latter to express the current i_1 flowing into the terminal at the left hand-side in Fig.135 yields

$$i_1 = \left(\frac{1}{\frac{1}{j\omega C_1} + R_1} + \frac{1}{\frac{1}{j\omega C} + R}\right) v_1 - \frac{1}{\frac{1}{j\omega C} + R} v_2. \quad (5.8)$$

Comparing (5.8) with (5.3), two of the y parameters are obtained as

$$y_{11} = \frac{1}{\frac{1}{j\omega C_1} + R_1} + \frac{1}{\frac{1}{j\omega C} + R} \quad (5.9)$$

$$y_{12} = -\frac{1}{\frac{1}{j\omega C} + R}. \quad (5.10)$$

Similarly, solving (5.5) for i, substituting it into (5.7) and rearranging the latter to express the current i_2 flowing into the terminal at the right hand-side in Fig.135 yields

$$i_2 = -\frac{1}{\frac{1}{j\omega C} + R} v_1 + \left(\frac{1}{\frac{1}{j\omega C_2} + R_2} + \frac{1}{\frac{1}{j\omega C} + R} \right) v_2. \tag{5.11}$$

Comparing (5.11) with (5.4), the other two y parameters are obtained as

$$y_{21} = y_{12} = -\frac{1}{\frac{1}{j\omega C} + R} \tag{5.12}$$

$$y_{22} = \frac{1}{\frac{1}{j\omega C_2} + R_2} + \frac{1}{\frac{1}{j\omega C} + R}. \tag{5.13}$$

The capacitance of the varactor C and its series loss resistance R can be calculated using (5.12) as

$$C = -\frac{1}{\omega \cdot \mathrm{Im}\left\{ -\frac{1}{y_{12}} \right\}} \tag{5.14}$$

$$R = \mathrm{Re}\left\{ -\frac{1}{y_{12}} \right\}. \tag{5.15}$$

Note that C_1 and R_1 can be calculated by adding y_{11} and y_{12} given in (5.9) and (5.10), respectively, as

$$C_1 = -\frac{1}{\omega \cdot \mathrm{Im}\left\{ \frac{1}{y_{11}+y_{12}} \right\}} \tag{5.16}$$

$$R_1 = \mathrm{Re}\left\{ \frac{1}{y_{11} + y_{12}} \right\}. \tag{5.17}$$

Similarly, C_2 and R_2 can be expressed by adding y_{21} and y_{22} given in (5.12) and (5.13), respectively, as

$$C_2 = -\frac{1}{\omega \cdot \operatorname{Im}\left\{\frac{1}{y_{21}+y_{22}}\right\}} \tag{5.18}$$

$$R_2 = \operatorname{Re}\left\{\frac{1}{y_{21}+y_{22}}\right\}. \tag{5.19}$$

A similar parameter extraction method was used for the integrated spiral inductors in which case the capacitor C in Fig.135 was naturally replaced with an inductor. The capacitance of the varactor computed at 2 GHz using (5.14) is plotted in Fig.134. The values of the other components were calculated to be higher than expected. Especially, the series resistive loss of the varactor was computed to be a large value yielding a quality factor of around 2 at 2 GHz for the varactor, which is consistent neither with the simulations nor with the filter measurements. The parameters extracted for the inductor were also found to be somehow unrealistic at frequencies higher than around 1 GHz. This is anticipated to be due to improper deembedding of the on-wafer probing pad parasitics from the measured parameters. In the case of the inductor, the rather excessive increase at the series loss of the inductor which reduces the quality factor to very low values inconsistently is attributed to improper grounding near the inductor layout. The conjecture is that the improper grounding (lack of good contacts around the inductor) could have left the inductor somehow floating, therefore yielding inconsistent parameter values in accordance with a similar equivalent circuit as in Fig.135, where the components at each port are in reference to the grounded substrate.

The frequency spectrum characterization of the filter was performed using an HP 8719ES Network Analyzer marked as "1" in Fig.80, whereas the dynamic range measurements were done using a Rohde&Schwarz FSEB 30 Spectrum Analyzer (marked as "4" in Fig.80), SMIQ 03 and HP8648C Signal Generators (marked as "2" and "6",

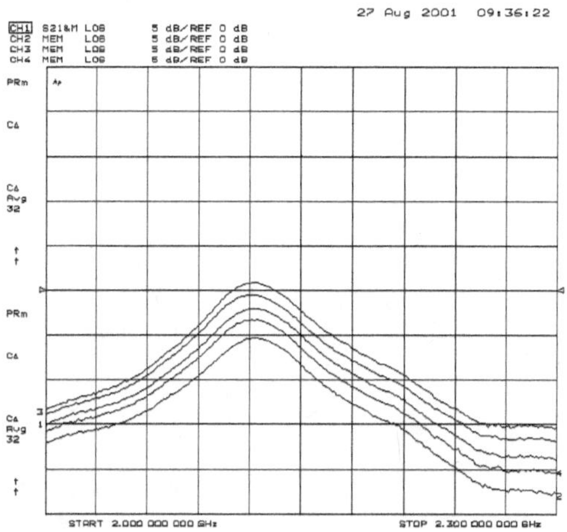

Fig. 136. Gain tuning of the RF filter with Q=40 and f_o=2.12GHz.

respectively in Fig.80) and an FSEK-3 noise figure measurement software (installed in the hard disk of the computer marked as "5" in Fig.80), which operates with the spectrum analyzer. It should also be noted that the noise figure measurements require a noise source to be connected after calibration to the input of the device under test which is connected to the spectrum analyzer. The power supply of the noise source is provided from the back panel of the spectrum analyzer. Some of the RF power measurements in the filter for the 1dB compression point characterization were taken with the Power Meter marked as "3" in Fig.80. The equipment marked as "8", "9" and "10" in the test setup are the general purpose multimeters and power supplies. A high-performance signal generator recently acquired in the laboratory is marked as "7" in Fig.80. All the performance figures measured, with the exception of the power consumption, are for the entire combination of the filter, output buffer and external passive circuitry supplying the bias and impedance matching.

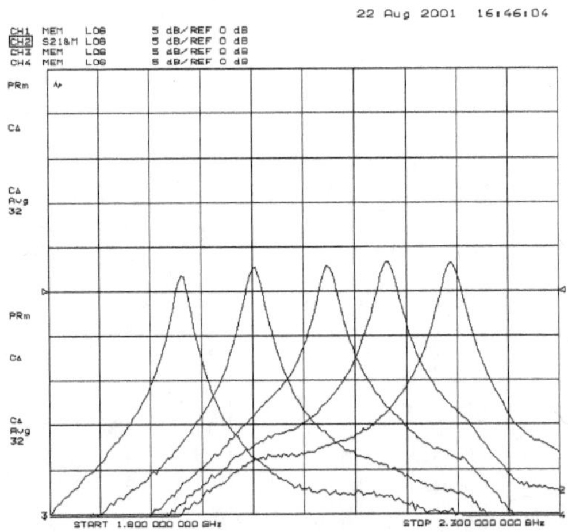

Fig. 137. Frequency tuning (f_o=1.93GHz - 2.19GHz) of the RF filter with $Q \approx$100.

The standalone measurements of the buffer show an attenuation of around 12dB at 2GHz. The gain programmability (around 2 octaves) of the biquad is shown in Fig.136. The varactors provide a frequency tuning range of 12.6% around 2.1GHz as depicted in Fig.137. The total current consumption of the circuit excluding the output buffer for a gain variation of around 6dB changes from 4mA to 7.25mA. In average, for a filter Q of 40, 80% of the total current is consumed by the negative conductance generator, whereas the remaining 20% is consumed by the input G_m stage. In the fabricated circuit, the devices $M_{1,2}$ and $M_{5,6}$ of Fig.47 were sized as $10 \times \frac{6\mu m}{0.4\mu m}$ and $12 \times \frac{6\mu m}{0.4\mu m}$, respectively. The Q tuning (more than 4 octaves) is illustrated in Fig.138. This much of variation in the Q of the filter requires around 5% of increase at the total current consumption in the measurements.

Simultaneous adjustments are made with both frequency and Q tuning to keep one constant while adjusting the other. To illustrate the programmability of the

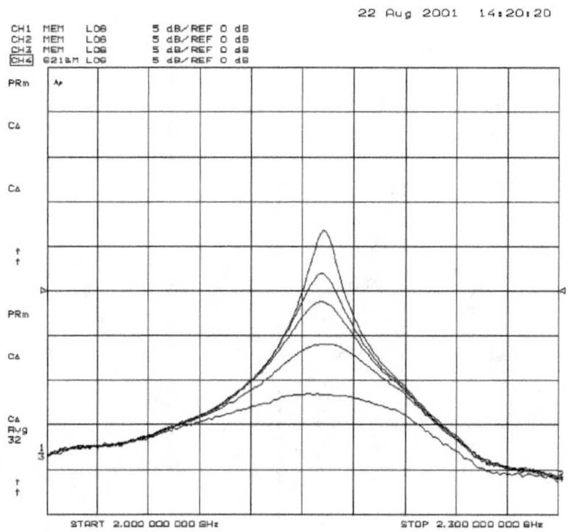

Fig. 138. Q tuning (Q=20 - 170) of the RF filter at f_o=2.16GHz.

filter, overlaid $|S_{21}|$ of ten different chips are plotted in Fig.139. In (a), the $|S_{21}|$ of the chips for the same bias settings are shown with the filter Q ranging from 20 to 80 and the center frequency of the filter varying between 2.14GHz - 2.18GHz. The plot in Fig.139(b) depicts the $|S_{21}|$ of the same ten chips after programmed for the same gain, Q and f_o using off-chip biases.

With V_{dd}=1.3V, P_{DC}=5.2mW, Q=40, f_o=2.19GHz, input 1 dB compression point, IIP_3 and the noise figure at the center frequency are measured as -30dBm, -18dBm and 26.8dB, respectively. Both the 1dB compression point and the third-order intercept point are referred to input, as shown in Figs.140-142. For the two-tone intermodulation distortion measurement, a two tone input signal including frequency components at 2.185GHz and 2.195GHz (5MHz higher and lower frequencies than the center frequency of the filter) at a filter Q setting of 40 is used. An Agilent 87303C power divider/combiner was used to generate the two tone signal at the input for the

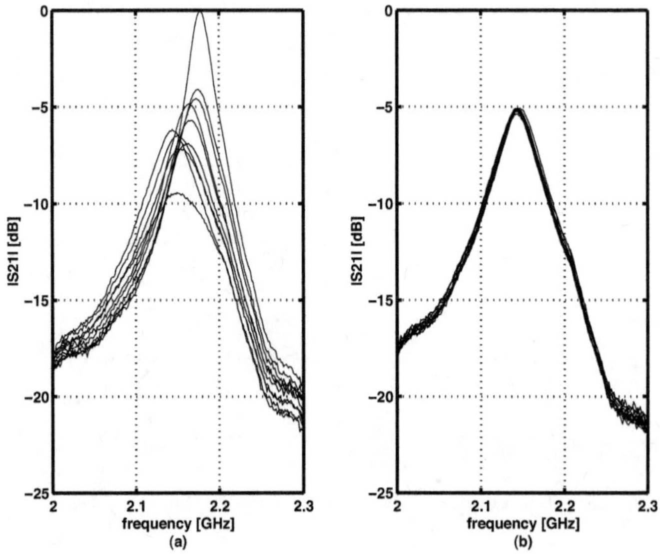

Fig. 139. Overlaid $|S_{21}|$ of ten different ICs (a) with the same bias settings and (b) after programmed for the same gain, Q=40 and f_o=2.14GHz.

test. The two tones at the input were kept at the same amplitude. The maximum power spectral density at the filter output and the noise floor at the filter input are extracted as -152dBm/Hz and -68dBm from the noise figure measurement, with noise integrated over the equivalent noise bandwidth of the filter. 1dB compression dynamic range of 35dB and spurious-free dynamic range (SFDR) of 31dB are obtained using the figures measured.

SFDR over the frequency tuning range for a constant filter Q of 40 was measured to change between 34-31dB as shown in Fig.143. A similar variation is observed in the IIP3 of the filter over the tuning range as Fig.144 depicts, thus the variation in the SFDR is attributed mainly to the nonlinearity of the varactor. Note that the

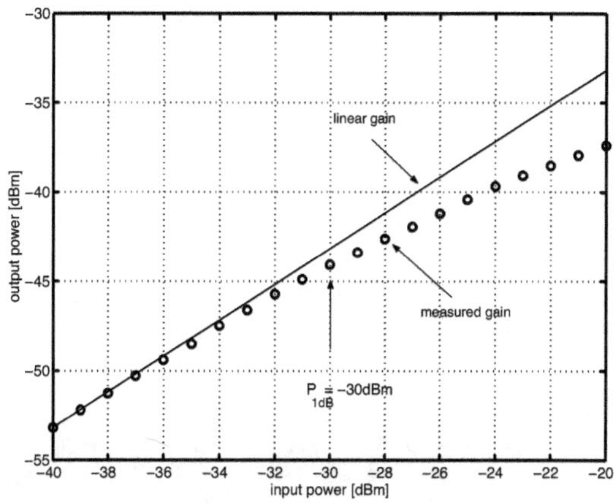

Fig. 140. 1dB compression point measurement of the RF filter at f_o=2.19GHz with Q=40.

variations in the noise figure and the input-referred noise floor over the tuning range shown in Figs. 145 and 146, respectively, do not exhibit a similar behavior as SFDR does. A variation in the IIP3 over the frequency tuning range is expectable because of the nonlinear capacitance-voltage transfer characteristic of the PMOS varactor illustrated in Fig.134.

The input g_m stage imposes trade-offs in the design of the filter between peak gain, power consumption, noise and linearity. Both simulations and measurements show that an optimization regarding the power consumption is possible by varying the g_m of the input stage (thus the peak gain of the filter), without substantially affecting the dynamic range. Fig.147 is an illustration of the trade-offs. In order to reduce the power consumption (by around 45%), the peak gain was experimentally reduced by 6dB, increasing the input-referred noise but improving the linearity to keep the dynamic range almost constant (a 1dB of decrease in the SFDR is observed).

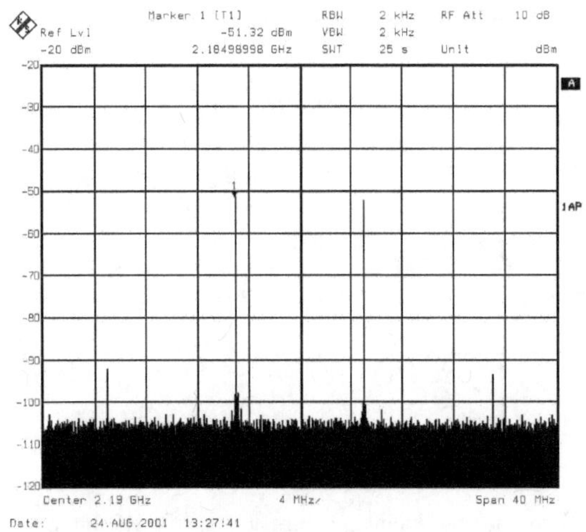

Fig. 141. Two tone measurement of the RF filter at f_o=2.19GHz with Q=40.

The measurements show that the filter operates with supply voltages ranging from 1.2V to 3V with the SFDR varying around 3dB for the same f_o and filter Q, as depicted in Fig.148. The simplicity of the structure allows for such a wide supply voltage range. As the supply voltage of the filter increases, a slight decrease in the noise figure is observed due to a slight increase at the peak gain because of the increased input transconductance. This is attributed to the increased drain source voltage of the tail transistor of the input differential pair, which in turn increases the tail current and thus the transconductance. The noise figure is reduced by around 1dB as the supply voltage is varied from 1.3V to 3V. Yet a more pronounced degradation in the linearity as the supply voltage is increased causes the reduction in the dynamic range of around 3dB as observed in Fig.148. The degradation in the linearity as the supply voltage gets higher can be attributed to the increased nonlinearity con-

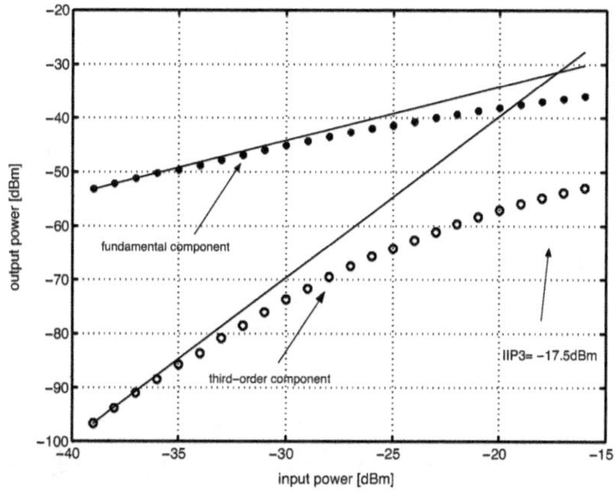

Fig. 142. Measured third-order intermodulation distortion of the RF filter at f_o=2.19GHz with Q=40.

tribution of the velocity saturation in the transistors, as well as the increased input stage transconductance, which makes the nonlinearity contributions of the negative conductance generator and the varactor more pronounced.

The strong dependance of the dynamic range to the quality factor of the filter is experimentally verified as shown in Fig.149 for a fixed V_{dd}=1.5V and center frequency of 2.17GHz. The lower dynamic range in the plot with Q=40 is due to the fact that the results shown here are of another chip on a more noisy printed-circuit board, which is irrelevant given the purpose of the plot. The main reason of the degradation in the dynamic range as the filter Q increases is the increased voltage swing across the nonlinear negative conductance circuit deteriorating the linearity of the filter as the variation of the 1dB compression point in Fig.150 shows. The match between the simulated and measured 1dB compression point depicted in the same figure is satisfactory.

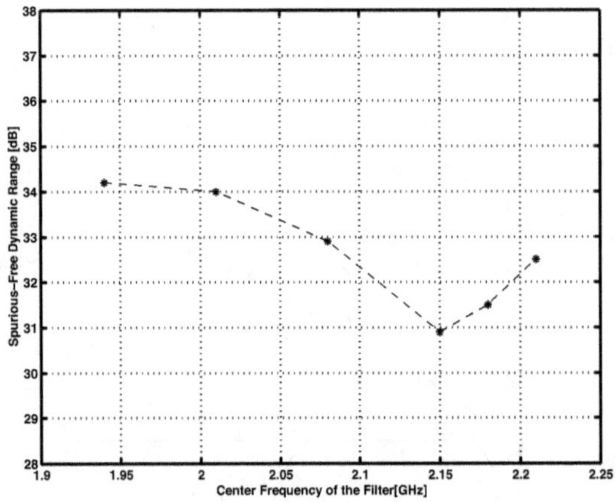

Fig. 143. Spurious-free dynamic range versus the center frequency of the filter for Q=40.

The bandwidth over which the output noise is to be integrated over to determine the minimum level in the dynamic range, depends on the application of the filter. It can be the channel bandwidth in a wireless communication standard if it is to be used in a receiver, [45] or for a more general characterization, the equivalent noise bandwidth of the filter transfer function can be used, [43]. 1dB compression point dynamic range and spurious-free dynamic range expressions are therefore parameterized with respect to the bandwidth B over which the noise is integrated. Substituting the measured values with V_{dd}=1.3V, I_{Vdd}=4mA, Q=40, f_o=2.19GHz, into (4.91) and (4.92) keeping B as a parameter, the dynamic range expressions are obtained as follows

$$
\begin{aligned}
DR_{1dB} &= 114.2 - 10\log B \\
SFDR &= \frac{2}{3}\left(126.2 - 10\log B\right). \qquad (5.20)
\end{aligned}
$$

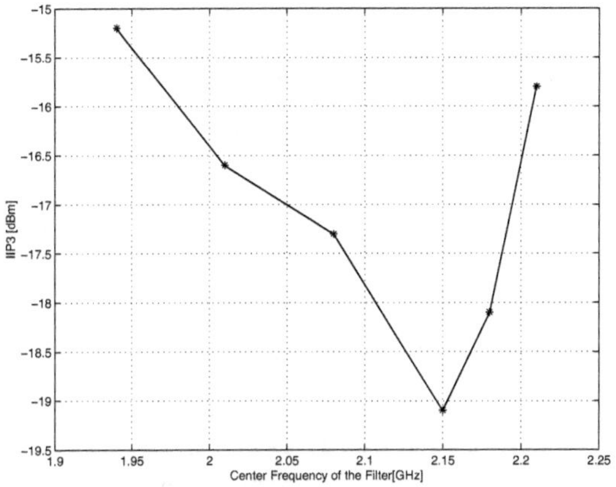

Fig. 144. Measured third-order intercept point versus the center frequency of the filter for Q=40.

Some dynamic range figures calculated using (5.20) based on the prototype measurements are tabulated in Table IX. Note that the bandwidth over which the noise is integrated, is taken as a parameter in the table.

Table IX. Calculated dynamic range figures based on the measurements with $V_{dd} = 1.3$V.

Noise Bandwidth	1dB Compression DR	Spurious-free DR
1MHz	54dB	44dB
5MHz	47dB	39dB
86MHz (Equivalent noise BW)	35dB	31dB

Table X shows a comparison for published CMOS, BiCMOS and Bipolar RF integrated bandpass filters in the literature. It should be noted that [93], [90] and [96] do not provide frequency tuning; [46] includes automatic tuning circuitry for

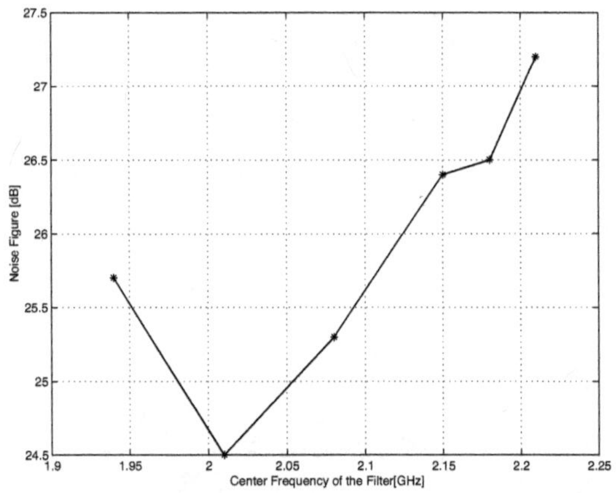

Fig. 145. Measured noise figure versus the center frequency of the filter for Q=40.

both frequency and Q; the SFDR figures given in [45], [96] and [98] are based on integrating the noise within 1MHz, 5MHz and 20MHz bandwidths, respectively. None of the dynamic range figures given in the reported papers take into consideration the additional 3dB in (4.89), which accounts for the difference between the minimum detectable signal and the noise floor. The comparison table demonstrates that the RF filter introduced uses low power supply voltage and low power consumption per pole and occupies small Silicon area per pole together with some of the other integrated Q-enhanced filters reported in the literature. A figure of merit (FOM) for evaluating the performances of the filters is defined as

$$FOM = \frac{N \cdot SFDR \cdot f_o \cdot Q}{P_D}. \tag{5.21}$$

where N denotes the number of poles in the filter. The FOM of the reported filters are also added in Table X at the last column. Despite the merits of the presented bandpass filter in this work mentioned above, due to its relatively low SFDR, its FOM

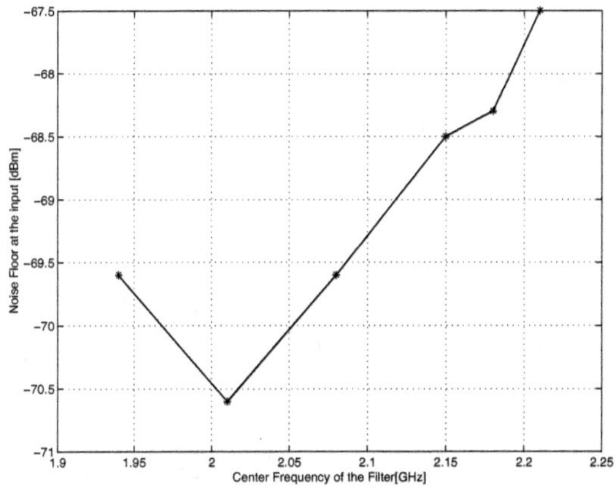

Fig. 146. Measured input-referred noise floor versus the center frequency of the filter for Q=40.

comes out as average among all the filters tabulated.

With the standalone buffer measurements, the filter linearity and noise performance figures like IIP_3 and Noise Figure can be deembedded. The deembedding is done using the measured results with V_{dd}=1.5V. The noise figure and the IIP_3 of the (filter + buffer) were measured as 26.9dB and -24.9dBm, respectively. On the other hand, the noise figure and the IIP_3 of the buffer alone were measured as 15dB and +13.5dBm, respectively. The gain of the filter itself at the center frequency was measured as around 15dB through deembedding the gain of the buffer from the measured gain of the (filter + buffer). Deembedding is done using the well known equations for the third-order intercept point and noise figure of cascaded systems, [7]. The noise figure of the filter is extracted from

$$F_{filter+buffer} \cong F_{filter} + \frac{F_{buffer} - 1}{A^2_{V,filter}} \tag{5.22}$$

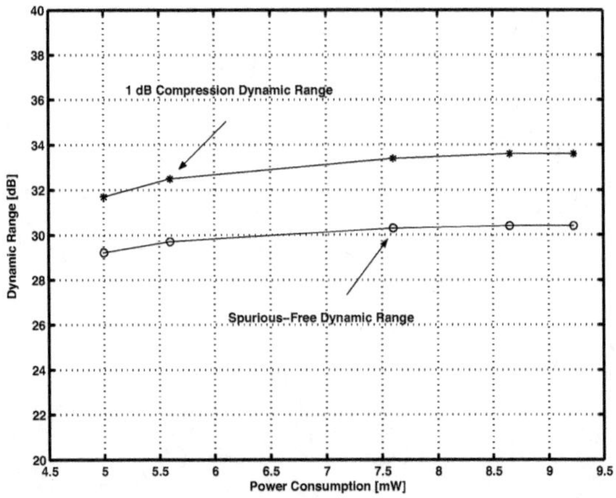

Fig. 147. Dynamic range versus the power consumption of the filter varied through the tail current of the input g_m stage, for f_o=2.17GHz and Q=40.

where F denotes the noise factor (Noise Figure= $10\log(F)$) and $A_{V,filter}$ denotes the voltage gain of the filter. The noise figure of the filter is solved to be 26.89dB, confirming that the measured noise figure of the (filter + buffer), which is 26.9dB, is quite an accurate representation of the noise figure of the filter. A similar deembedding is done regarding the input third-order intercept point of the filter. The IIP_3 of the filter is extracted from

$$\frac{1}{A^2_{IIP3,filter+buffer}} \cong \frac{1}{A^2_{IIP3,filter}} + \frac{A^2_{V,filter}}{A^2_{IIP3,buffer}} \tag{5.23}$$

where A^2_{IIP3} denotes the input third-order intercept point in $[V^2]$ in a 50Ω system. The input third-order intercept point of the filter is solved to be -24.88dBm using (5.23), confirming that the measured input third-order intercept point of the (filter + buffer), which is -24.9dBm, is quite an accurate representation of the third-order intercept point of the filter. Note that

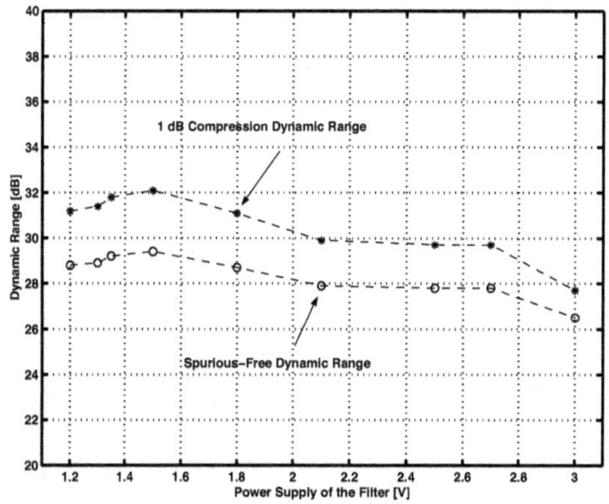

Fig. 148. Dynamic range versus the supply voltage of the filter for Q=40 and f_o=2.17GHz.

$$IIP_3[dBm] \equiv 10\log\left(\frac{A^2_{IIP3}/50}{1mW}\right) \cdot \tag{5.24}$$

The measured noise figures of the standalone buffer and the (filter + buffer) combination, 1-dB compression point and the third-order intercept point of the standalone buffer are shown in Figs.151-154.

The measured performance of the filter should be evaluated in the context of a wireless communication standard specification so that the bottlenecks of the design can be better assessed, the characteristics of the circuit to be improved in a future design can be determined in a more concrete way. A comparison between the rather stringent specifications of an image reject filter in a Multi Standard Receiver for DCS/UMTS and the measured performance of the filter presented in this subsection is tabulated in Table XI. The standard specifications in Table XI were put together by a colleague in the Analog&Mixed-Signal Design Group in Texas A&M University,

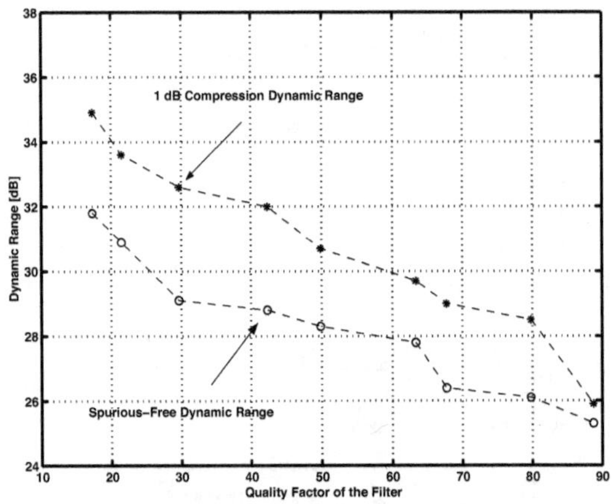

Fig. 149. Dynamic range versus the filter Q with f_o=2.17GHz and V_{dd}=1.5V.

taken into account the specifications of DCS and UMTS Standards available in the literature.

It is seen in Table XI that the frequency tuning range of the filter should be increased. A digitally programmable linear capacitor array can be used in case the technology the filter is to be designed with allows for linear capacitors such as poly-poly capacitors. Providing the frequency tuning with a programmable linear capacitor array rather than a PMOS varactor would also improve the linearity of the filter. Recall from the nonlinearity analyses presented in Chapter IV that the PMOS varactor is among the dominant nonlinearity contributors in the filter. The power gain specification of the filter can be met as the peak gain of the filter structure is programmable with an off-chip bias. The noise figure of the filter on the other hand should be improved. Using a spiral inductor with a higher quality factor would considerably enhance both the noise and linearity performance of the filter as depicted in Fig.51 and demonstrated in the noise and nonlinearity analyses covered in Chapter IV. It should

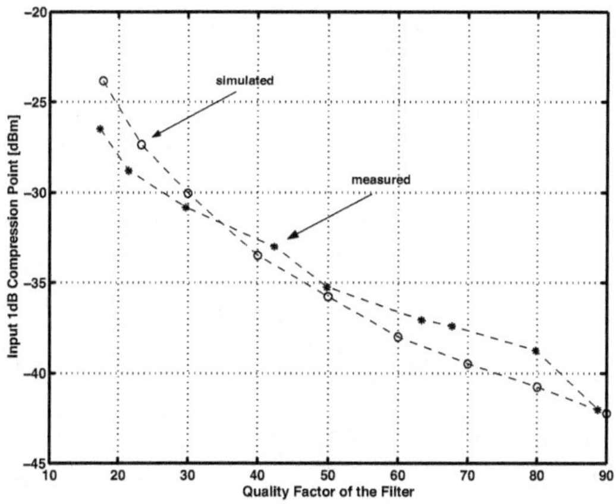

Fig. 150. Comparison between the measured and simulated 1dB compression point versus the filter quality factor with f_o=2.17GHz and V_{dd}=1.5V.

be repeated that the inductors in the filter are the most dominant noise contributors and using inductors with higher quality factor would reduce the noise substantially.

A strong emphasis should be put to improve the linearity of the design and the best approach to be taken is to linearize the negative conductance generator which is the most dominant nonlinearity contributor. Source-degeneration resistors can be used in the cross-coupled differential pairs with some penalty in the power consumption. Note that using an LC tank with higher quality factor would largely compensate for the increase in the power consumption that would result from the use of source-degeneration resistors.

It is seen in Table XI that no improvement in the image rejection is observed between 1546 MHz and 1256 MHz in the filter. In order to accurately determine the image rejection in the filter, a broadband impedance matching should be achieved in the PCB the filter is characterized on. Namely, low and flat S_{11} and S_{22} over the

Table X. Performance comparison of CMOS, BiCMOS and Bipolar integrated RF filters in the literature.

Ref.	Tech.	f_o	Bandwidth	V_{dd}	P_D/pole	Area/pole	SFDR	FOM
[43]	Bipolar	1.8GHz	51.4MHz	2.8V	12.2mW	0.2mm^2	30dB	127dB
[93]	Bipolar	1GHz	25MHz	5V	34mW	0.3mm^2	36dB	127dB
[90]	BiCMOS	750MHz	37.5MHz	5V	40mW	0.3mm^2	25dB	111dB
[46]	BiCMOS	1.9GHz	150MHz	2.7V	12.15mW	1.79mm^2	49dB	142dB
[45]	CMOS	850MHz	18MHz	2.7V	52mW	0.5mm^2	55dB	144dB
[95]	CMOS	850MHz	28.3MHz	2V	22.9mW	0.32mm^2	28dB	119dB
[96]	CMOS	2.14GHz	60MHz	2.5V	2.92mW	0.59mm^2	55dB	159dB
[97]	CMOS	1.85GHz	80MHz	2.7V	10.8mW	0.0375mm^2	35dB	131dB
[98]	CMOS	4.7GHz	18.8MHz	1.5V	0.6mW	0.2mm^2	35dB	158dB
[102]	CMOS	5GHz	475MHz	1.2V	1.44mW	0.15mm^2	57dB	163dB
This work	CMOS	2.19GHz	53.8MHz	1.3V	2.6mW	0.05mm^2	31dB	136dB

frequency band of interest should be obtained, so that the impedance matching do not interfere with the frequency response of the filter. The impedance matching on the PCB the RF filter was characterized with lacked such a broadband and good matching. The inconsistency in the measured image rejection is attributed to this poor impedance matching. It is suggested that two Biquads such as the filter presented be cascaded to form a fourth-order filter to improve the image rejection to the levels required in the Multi Standard Receiver specification. Although its measured performance is short of meeting the stringent specifications tabulated in Table XI, the filter still holds its merits of being operable with a relatively low power supply voltage of $V_{dd} = 1.3V$, consuming low power per pole and occupying small silicon

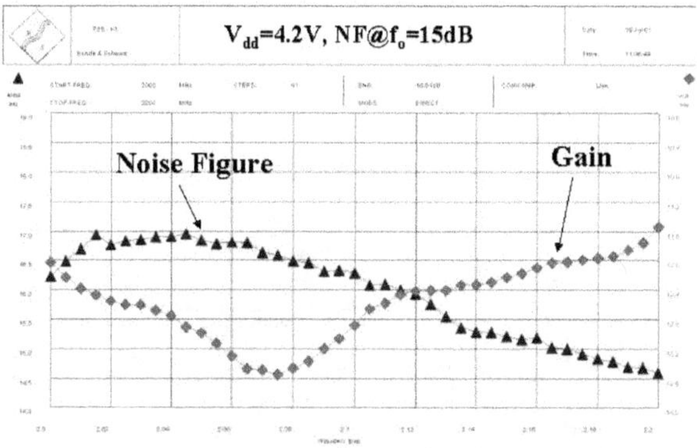

Fig. 151. Measured noise figure of the standalone output buffer of the filter.

area per pole among the similar filters reported in the literature, as demonstrated in Table X.

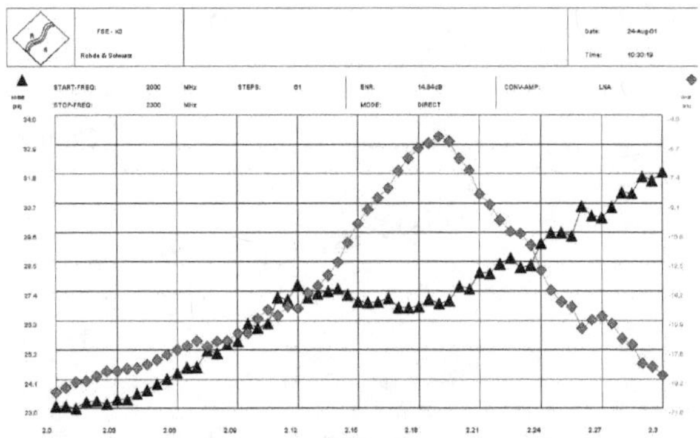

Fig. 152. Measured noise figure of the (filter + buffer) with $V_{dd} = 1.3V$ and $Q = 40$.

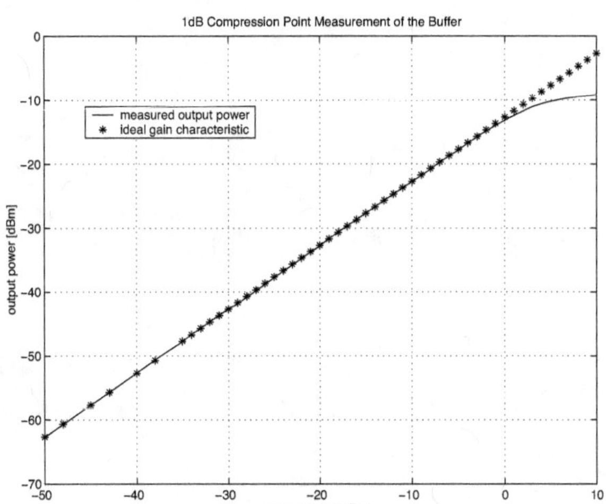

Fig. 153. Measured 1-dB compression point of the standalone output buffer of the filter (P_{1dB}=2dBm).

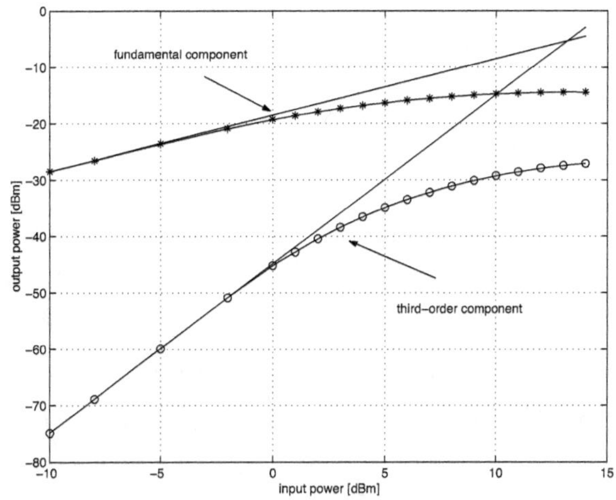

Fig. 154. Measured third-order intercept point of the standalone output buffer of the filter (IIP_3=13.5dBm).

Table XI. Comparison between the specifications of an image reject filter in a multi standard receiver for DCS/UMTS and the measured performance of the RF filter.

Parameter	Standard Specification	This work
3 dB BW	1800-2180 MHz	1930-2190 MHz
Power Gain	0-3	can be adjusted
Noise Figure	15 dB	26.8 dB
Input 1 dB Compression Point	-5 dBm	-30 dBm
Input 3rd-Order Intercept Point	5 dBm	-18 dBm
Image Rejection at 1546 MHz	30 dB	25 dB
Image Rejection at 1256 MHz	45 dB	25 dB

CHAPTER VI

CONCLUSIONS AND FUTURE WORK

Negative conductance VCOs with fully integrated resonators in Si technologies show promising results as the quality of the integrated resonator passive elements are improved by the designers and process developers. Recent improvements in the phase noise theory enable the designers to correctly identify the component contributions for optimization in the design process. The design trade-offs of negative-conductance VCO structures in CMOS and BiCMOS technologies together with experimental results of two fully integrated CMOS VCOs which use NMOS cross-coupled pairs to generate negative conductance that compensate for the LC tank losses are presented. The VCOs oscillate at 2.2GHz.

Design considerations involved in a symmetric linearized CMOS VCO architecture with integrated tank elements are introduced. The impact of the linearizing source-degeneration resistors on the phase noise is demonstrated. The symmetry and balanced nature of the proposed architecture provides additional design degrees of freedom. The source-degeneration resistors allow the designer to trade the close-in phase noise with the phase noise at large offsets from the carrier and with the frequency tuning range for a given bias current. Discussions on the practical limitations of the integrated LC tank components of the VCO are provided with emphasize on the limiting effect of the well resistance of the bulk-tuned PMOS capacitors used as varactors, on the frequency tuning. Calculated, SpectreRF simulated and measured phase noise results within the offset frequency range of 1kHz to 3MHz from the carrier are in good agreement.

Design issues in a dual-modulus divide by 32/33 prescaler in BiCMOS are presented. The phase noise contributors in the circuit determined through simulations are investigated and some guidelines are provided to improve the noise performance. The impact of the integrated bias network on the phase noise of the prescaler is demonstrated together with the trade-off between the power consumption and the phase noise. The match between the simulated and the measured phase noise is within 5dB for offsets ranging from 1kHz to 1MHz.

A 2.1GHz, CMOS fully integrated second-order Q-enhancement LC Bandpass filter with tunable center frequency, peak gain and quality factor is introduced. The filter uses a resonator built with spiral inductors and inversion-mode PMOS capacitors which provide frequency tuning. The dynamic range issues are emphasized in the discussions. Noise and nonlinearity analyses are presented to determine the design trade-offs involved in the circuit. Simplified analytical expressions for the nonlinearity contributions of the negative conductance generator, the varactor and the input G_m stage are provided. An expression for the noise factor of the filter is also derived. The expressions which are verified with simulations demonstrate the design trade-offs. Measured frequency tuning range of the filter around 2.1GHz is 13%. Spiral inductors with Q of 2 at 2.1GHz limit the spurious-free dynamic range at a minimum of 31dB within the frequency tuning range. The filter implemented in a 0.35 μm standard CMOS technology sinks 4mA from a 1.3V supply providing a filter quality factor of 40 at 2.19GHz with a 1dB compression dynamic range of 35dB. Simulations show that an improvement of around 15dB is possible with spiral inductors that have a quality factor of 20. The RF filter introduced uses low power supply voltage and low power consumption per pole and occupies small silicon area per pole when compared to the similar filters reported in the literature. The measured performance of the filter presented shows the feasibility of designing low voltage, low power fully-integrated programmable bandpass filters in mainstream CMOS technologies at frequencies higher than 2GHz.

An important conclusion to be pointed out from the overall research is the impact of the quality factor of the integrated spiral inductors on the performance of both the negative conductance LC VCOs and the Q-enhancement LC filters. The lack of reliable models for the varactors that provide the frequency tuning in the aforementioned building blocks impair the design process in the sense that some important effects cannot be captured in the simulations. The reduction at the frequency tuning range due to the well resistance of the bulk-tuned PMOS varactor which was not adequately modeled at the simulation phase is a typical example of this. Redesigning the RF Bandpass filter in CMOS with adequately modeled higher quality passive components, especially spiral inductors would provide a better dynamic range performance. The design trade-offs demonstrated in the noise and nonlinearity analyses in Chapter IV should be incorporated carefully into the design process. As mentioned before, it is suggested that two Biquads such as the filter presented be cascaded to form a fourth-order filter to improve the image rejection to the levels required in Wireless Communication System Standards. The possibilities of embedding automatic tuning circuitry for frequency, Q and peak gain into the filter should be investigated thoroughly in order to enhance the practicality of the circuit for industrial applications.

REFERENCES

[1] F. Dülger, "Fully integrated building blocks for wireless communication transceiver front-ends on silicon," Ph.D. dissertation, Texas A&M University, College Station, 2002.

[2] F. Dülger and E. Sánchez-Sinencio, "Fully-integrated LC VCOs at RF on silicon," in *Proceedings of the Southwest Symposium on Mixed-Signal Design*, Austin, February 2001, pp. 13–18.

[3] F. Dülger and E. Sánchez-Sinencio, "Design trade-offs of a symmetric linearized CMOS LC VCO," in *Proceedings of the IEEE International Symposium on Circuits and Systems (ISCAS)*, Phoenix, May 2002, pp. IV–397 - IV–400.

[4] F. Dülger, E. Sánchez-Sinencio, and A. Bellaouar, "Design considerations in a BiCMOS dual-modulus prescaler," in *Digest of papers, IEEE Radio Frequency Integrated Circuits (RFIC) Symposium*, Seattle, June 2002, pp. 177–180.

[5] F. Dülger, E. Sánchez-Sinencio, and J. S. Martínez, "A 2.1GHz 1.3V 5mW programmable Q-enhancement LC bandpass biquad in 0.35μm CMOS," in *Proceedings of the IEEE Custom Integrated Circuits Conference*, Orlando, May 2002, pp. 283–286.

[6] F. Dülger, E. Sánchez-Sinencio, and J. S. Martínez, "A 1.3V 5mW fully integrated tunable bandpass filter at 2.1GHz in 0.35μm CMOS," *IEEE Journal of Solid-State Circuits*, vol. 38, no. 6, pp. 918–928, June 2003.

[7] B. Razavi, *RF Microelectronics*, Prentice Hall PTR, New Jersey, 1998.

[8] P. Leroux, J. Janssens, and M. Steyaert, "A 0.8 dB NF ESD-protected 9 mW CMOS LNA," in *International Solid-State Circuits Conference Digest of Technical Papers*, San Francisco, February 2001, pp. 410–411.

[9] H. Hashemi and A. Hajimiri, "Concurrent multiband low-noise amplifiers-theory, design, and applications," *IEEE Transactions on Microwave Theory and Techniques*, vol. 50, no. 1, pp. 288–301, January 2002.

[10] L. Yi, M. Obrecht, and T. Manku, "RF noise characterization of MOS devices for LNA design using a physical-based quasi-3-D approach," *IEEE Transactions on Circuits and Systems-II*, vol. 48, no. 10, pp. 972–984, October 2001.

[11] G. Gramegna, M. Paparo, P. G. Erratico, and P. De Vita, "A sub-1-dB NF ±2.3-kV ESD-protected 900-MHz CMOS LNA," *IEEE Journal of Solid-State Circuits*, vol. 36, no. 7, pp. 1010–1017, July 2001.

[12] F. Gatta, E. Sacchi, F. Svelto, P. Vilmercati, and R. Castello, "A 2-dB noise figure 900-MHz differential CMOS LNA," *IEEE Journal of Solid-State Circuits*, vol. 36, no. 10, pp. 1444–1452, October 2001.

[13] B. Floyd and D. Öziş, "Low noise amplifier comparison at 2 GHz in 0.25μm and 0.18μm RF-CMOS and SiGe BiCMOS," in *Digest of papers, IEEE Radio Frequency Integrated Circuits (RFIC) Symposium*, Fort Worth, June 2004, pp. 185–188.

[14] B. G. Perumana, J. H. C. Zhan, S. S. Taylor, B. R. Carlton, and J. Laskar, "Resistive-feedback CMOS low noise amplifiers for multiband applications," *IEEE Transactions on Microwave Theory and Techniques*, vol. 56, no. 5, pp. 1218–1225, May 2008.

[15] J. A. Weldon, R. S. Narayanaswami, J.C. Rudell, L. Li, M. Otsuka, S. Dedieu, T. Luns, T. King-Chun, L. Cheol-Woong, and P. R. Gray, "A 1.75-GHz highly integrated narrow-band CMOS transmitter with harmonic-rejection mixers," *IEEE Journal of Solid-State Circuits*, vol. 36, no. 12, pp. 2003–2015, December 2001.

[16] D. Coffing and E. Main, "Effects of offsets on bipolar integrated circuit mixer even-order distortion terms," *IEEE Transactions on Microwave Theory and Techniques*, vol. 49, no. 1, pp. 23–30, January 2001.

[17] T. Melly, A. S. Porret, C. C. Enz, and E. A. Vittoz, "An analysis of flicker noise rejection in low-power and low-voltage CMOS mixers," *IEEE Journal of Solid-State Circuits*, vol. 36, no. 1, pp. 102–109, January 2001.

[18] T. Hornak, K. L. Knudsen, A. Z. Grzegorek, K. A. Nishimura, and W. J. McFarland, "An image-rejecting mixer and vector filter with 55-dB image rejection over process, temperature, and transistor mismatch," *IEEE Journal of Solid-State Circuits*, vol. 36, no. 1, pp. 23–33, January 2001.

[19] H. D. Wohlmuth and W. Simbürger, "A high-IP3 RF receiver chip set for mobile radio base stations up to 2GHz," *IEEE Journal of Solid-State Circuits*, vol. 36, no. 7, pp. 1132–1137, July 2001.

[20] K. J. Koh, M. Y. Park, C. S. Kim, and H. K. Yu, "Subharmonically pumped CMOS frequency conversion (up and down) circuits for 2-GHz WCDMA direct-conversion transceiver," *IEEE Journal of Solid-State Circuits*, vol. 39, no. 6, pp. 871–884, June 2004.

[21] J. Park, C. H. Lee, B. S. Kim, and J. Laskar, "Design and analysis of low flicker noise CMOS mixers for direct conversion receivers," *IEEE Transactions*

on Microwave Theory and Techniques, vol. 54, no. 12, pp. 4372–4380, December 2006.

[22] K. L. R. Mertens and M. S. J. Steyaert, "A 700-MHz 1-W fully differential CMOS class-E power amplifier," *IEEE Journal of Solid-State Circuits*, vol. 37, no. 2, pp. 137–141, February 2002.

[23] I. Aoki, S. D. Kee, D. B. Rutledge, and A. Hajimiri, "Fully integrated CMOS power amplifier design using the distributed active-transformer architecture," *IEEE Journal of Solid-State Circuits*, vol. 37, no. 3, pp. 371–383, March 2002.

[24] G. Hau, T. B. Nishimura, and N. Iwata, "A highly efficient linearized wide-band CDMA handset power amplifier based on predistortion under various bias conditions," *IEEE Transactions on Microwave Theory and Techniques*, vol. 49, no. 6, pp. 1194–1201, June 2001.

[25] C. Yoo and Q. Huang, "A common-gate switched 0.9-W class-E power amplifier with 41% PAE in 0.25μm CMOS," *IEEE Journal of Solid-State Circuits*, vol. 36, no. 5, pp. 823–830, May 2001.

[26] R. Gupta, B. M. Ballweber, and D. J. Allstot, "Design and optimization of CMOS RF power amplifiers," *IEEE Journal of Solid-State Circuits*, vol. 36, no. 2, pp. 166–175, February 2001.

[27] P. Reynaert and M. S. J. Steyaert, "A 1.75-GHz polar modulated CMOS RF power amplifier for GSM-EDGE," *IEEE Journal of Solid-State Circuits*, vol. 40, no. 12, pp. 2598–2608, December 2005.

[28] Y. S. Jeon, J. Cha, and S. Nam, "High efficiency power amplifier using novel

dynamic bias switching," *IEEE Transactions on Microwave Theory and Techniques*, vol. 55, no. 4, pp. 690–696, April 2007.

[29] G. Chien and P. R. Gray, "A 900-MHz local oscillator using a DLL-based frequency multiplier technique for PCS applications," *IEEE Journal of Solid-State Circuits*, vol. 35, pp. 1996–1999, December 2000.

[30] A. Hajimiri and T. H. Lee, *The Design of Low Noise Oscillators*, Kluwer Academic Publishers, Boston, 1999.

[31] J. Craninckx and M. Steyaert, *Wireless CMOS Frequency Synthesizer Design*, Kluwer Academic Publishers, Dordrecht, 1998.

[32] J. Craninckx and M. Steyaert, "A 1.75-GHz/3V dual-modulus divide-by-128/129 prescaler in 0.7-μm CMOS," *IEEE Journal of Solid-State Circuits*, vol. 31, no. 7, pp. 890–897, July 1996.

[33] C. S. Vaucher, I. Ferencic, M. Locher, S. Sedvallson, U. Voegeli, and Z. Wang, "A family of low-power truly modular programmable dividers in standard 0.35-μm CMOS technology," *IEEE Journal of Solid-State Circuits*, vol. 35, no. 7, pp. 1039–1045, July 2000.

[34] J. Craninckx and M. Steyaert, "A fully integrated CMOS DCS-1800 frequency synthesizer," *IEEE Journal of Solid-State Circuits*, vol. 33, no. 12, pp. 2054–2065, December 1998.

[35] W. S. T. Yan and H. C. Luong, "A 2V 900-MHz monolithic CMOS dual-loop frequency synthesizer for GSM receivers," *IEEE Journal of Solid-State Circuits*, vol. 36, pp. 204–216, February 2001.

[36] B. Jansen, K. Negus, and D. Lee, "Silicon bipolar VCO family for 1.1 to 2.2 GHz with fully integrated tank and tuning circuits," in *International Solid-State Circuits Conference Digest of Technical Papers*, San Francisco, February 1997, pp. 392–393.

[37] M. A. Margarit, J. L. Tham, R. G. Meyer, and M. J. Deen, "A low noise, low power VCO with automatic amplitude control for wireless applications," *IEEE Journal of Solid-State Circuits*, vol. 34, no. 6, pp. 761–771, June 1999.

[38] A. M. Niknejad, J. L. Tham, and R. G. Meyer, "Fully integrated low phase noise bipolar differential VCOs at 2.9 and 4.4 GHz," in *Proceedings of European Solid-State Circuits Conference*, 1999, pp. 198–201.

[39] H. Wang, "A 9.8GHz back-gate tuned VCO in 0.35μm CMOS," in *International Solid-State Circuits Conference Digest of Technical Papers*, San Francisco, February 1999, pp. 406–407.

[40] B. D. Muer, M. Borremans, M. Steyaert, and G. L. Puma, "A 2 GHz low phase noise integrated LC VCO set with flicker noise upconversion minimization," *IEEE Journal of Solid-State Circuits*, vol. 35, no. 7, pp. 1034–1038, July 2000.

[41] T. H. Lee and A. Hajimiri, "Oscillator phase noise: a tutorial," *IEEE Journal of Solid-State Circuits*, vol. 35, pp. 326–336, March 2000.

[42] H. Knapp, W. Wilhelm, and M. Wurzer, "A low-power 15-GHz frequency divider in a 0.8μm silicon bipolar technology," *IEEE Transactions on Microwave Theory and Techniques*, vol. 48, no. 2, pp. 205–208, February 2000.

[43] S. Pipilos, Y. P. Tsividis, J. Fenk, and Y. Papananos, "A Si 1.8 GHz RLC filter with tunable center frequency and quality factor," *IEEE Journal of Solid-State*

Circuits, vol. 31, no. 10, pp. 1517–1525, October 1996.

[44] J. A. Macedo and M. A. Copeland, "A 1.9-GHz silicon receiver with monolithic image filtering," *IEEE Journal of Solid-State Circuits*, vol. 33, no. 3, pp. 378–386, March 1998.

[45] W. B. Kuhn, N. K. Yanduru, and A. S. Wyszynski, "Q-enhanced LC bandpass filters for integrated wireless applications," *IEEE Transactions on Microwave Theory and Techniques*, vol. 46, no. 12, pp. 2577–2586, December 1998.

[46] D. Li and Y. Tsividis, "A 1.9GHz Si active LC filter with on-chip automatic tuning," in *International Solid-State Circuits Conference Digest of Technical Papers*, San Francisco, February 2001, pp. 368–369.

[47] E. Hegazi, H. Sjöland, and A. Abidi, "A filtering technique to lower LC oscillator phase noise," *IEEE Journal of Solid-State Circuits*, vol. 36, no. 12, pp. 1921–1930, December 2001.

[48] R. Aparicio and A. Hajimiri, "A CMOS differential noise-shifting colpitts VCO," in *International Solid-State Circuits Conference Digest of Technical Papers*, San Francisco, February 2002, pp. 288–289.

[49] M. Tiebout, H. D. Wohlmuth, and W. Simbürger, "A 1V 51GHz fully-integrated VCO in 0.12μm CMOS," in *International Solid-State Circuits Conference Digest of Technical Papers*, San Francisco, February 2002, pp. 300–301.

[50] A. H. Mostafa, M. N. El-Gamal, and R. A. Rafla, "A sub-1V 4-GHz CMOS VCO and a 12.5-GHz oscillator for low-voltage and high-frequency applications," *IEEE Transactions on Circuits and Systems-II*, vol. 48, no. 10, pp. 919–926, October 2001.

[51] A. R. Kral, "A 2.4GHz frequency synthesizer in 0.6μm CMOS," M.S. thesis, UCLA, Los Angeles, 1998.

[52] M. D. M. Hershenson, A. Hajimiri, S. S. Mohan, S. P. Boyd, and T. H. Lee, "Design and optimization of LC oscillators," in *IEEE/ACM International Conference on Computer-Aided Design Digest of Technical Papers*, San Jose, CA, November 1999, pp. 65–69.

[53] D. Ham and A. Hajimiri, "Design and optimization of a low noise 2.4GHz CMOS VCO with integrated LC tank and MOSCAP tuning," in *Proceedings of International Symposium on Circuits and Systems*, May 2000.

[54] D. Ham and A. Hajimiri, "Concepts and methods in optimization of integrated LC vcos," *IEEE Journal of Solid-State Circuits*, vol. 36, no. 6, pp. 896–909, June 2001.

[55] A. Hajimiri and T. H. Lee, "Design issues in CMOS differential LC oscillators," *IEEE Journal of Solid-State Circuits*, vol. 34, no. 5, pp. 717–724, May 1999.

[56] H. Wang, A. Hajimiri, and T. H. Lee, "Correspondence:comments on "design issues in CMOS differential LC oscillator," *IEEE Journal of Solid-State Circuits*, vol. 35, no. 2, pp. 286–287, February 2000.

[57] T. Wakimoto and S. Konaka, "A 1.9GHz Si bipolar quadrature VCO with fully-integrated LC tank," in *Symposium of VLSI Circuits Digest of Technical Papers*, June 1998, pp. 30–31.

[58] H. Wang, "A solution for minimizing phase noise in low-power resonator-based oscillators," in *Proceedings of International Symposium on Circuits and Systems*, May 2000.

[59] D. B. Leeson, "A simple model of feedback oscillator noise spectrum," *Proceedings of IEEE*, vol. 54, pp. 329–330, February 1966.

[60] J. N. Burghartz, M. Soyuer, and K. A. Jenkins et al., "Integrated RF components in a SiGe bipolar technology," *IEEE Journal of Solid-State Circuits*, vol. 32, no. 9, pp. 1440–1445, September 1997.

[61] T. Soorapanth, C. P. Yue, and D. K. Shaeffer et al., "Analysis and optimization of accumulation-mode varactor for RF IC's," in *Symposium of VLSI Circuits Digest of Technical Papers*, June 1998, pp. 32–33.

[62] A. S. Porret, T. Melly, and C. C. Enz, "Design of high-Q varactors for low-power wireless applications using a standard CMOS process," in *Proceedings of Custom Integrated Circuits Conference*, Santa Clara, CA, May 1998, pp. 641–644.

[63] P. Andreani, "A comparison between two 1.8GHz CMOS VCO's tuned by different varactors," in *Proceedings of European Solid-State Circuits Conference*, September 1998, pp. 380–383.

[64] A. S. Porret, T. Melly, C. C. Enz, and E. A. Vittoz, "Design of high-Q varactors for low-power wireless applications using a standard CMOS process," *IEEE Journal of Solid-State Circuits*, vol. 35, no. 3, pp. 337–345, March 2000.

[65] W. M. Y. Wong, P. S. Hui, Z. Chen, and K. Shen, "A wide tuning range gated varactor," *IEEE Journal of Solid-State Circuits*, vol. 35, no. 5, pp. 773–779, May 2000.

[66] P. Andreani and S. Mattisson, "On the use of MOS varactors in RF VCO's," *IEEE Journal of Solid-State Circuits*, vol. 35, no. 6, pp. 905–910, June 2000.

[67] S. M. Sze, *Physics of Semiconductor Devices*, Wiley, New York, NY, 1981.

[68] Y. P. Tsividis, *Operation and Modeling of the MOS Transistor*, McGraw-Hill, New York, NY, 2nd edition, 1999.

[69] R. H. Kingston and S. F. Neustadter, "Calculation of the space charge, electric field and free carrier concentration at the surface of a semiconductor," *Journal of Applied Physics*, vol. 26, no. 6, pp. 718–720, June 1955.

[70] K. A. Kouznetsov and R. G. Meyer, "Phase noise in LC oscillators," *IEEE Journal of Solid-State Circuits*, vol. 35, no. 8, pp. 1244–1248, August 2000.

[71] H. Wu and A. Hajimiri, "A 19GHz 0.5mW 0.35μm CMOS frequency divider with shunt-peaking locking-range enhancement," in *International Solid-State Circuits Conference Digest of Technical Papers*, San Francisco, February 2001, pp. 412–413.

[72] UC Berkeley, Berkeley, CA, *BSIM3v3 Manual*, 1999.

[73] T. Seneff, L. McKay, K. Sakamoto, and N. Tracht, "A sub-1 mA 1.5-GHz silicon bipolar dual modulus prescaler," *IEEE Journal of Solid-State Circuits*, vol. 29, no. 10, pp. 1206–1211, October 1994.

[74] N. Foroudi and T. A. Kwasniewski, "CMOS high-speed dual-modulus frequency divider for RF frequency synthesis," *IEEE Journal of Solid-State Circuits*, vol. 30, no. 2, pp. 93–100, February 1995.

[75] H. Wang, "A 1.8V 3mW 16.8GHz frequency divider in 0.25μm CMOS," in *International Solid-State Circuits Conference Digest of Technical Papers*, San Francisco, February 2000, pp. 196–197.

[76] B. Razavi, K. F. Lee, and R. H. Yan, "Design of high-speed, low-power frequency dividers and phase-locked loops in deep submicron CMOS," *IEEE Journal of Solid-State Circuits*, vol. 30, no. 2, pp. 101–109, February 1995.

[77] H. R. Rategh and T. H. Lee, "Superharmonic injection locked oscillators as low power frequency dividers," in *Symposium of VLSI Circuits Digest of Technical Papers*, June 1998, pp. 132–135.

[78] H. R. Rategh and T. H. Lee, "Superharmonic injection-locked frequency dividers," *IEEE Journal of Solid-State Circuits*, vol. 34, no. 6, pp. 813–821, June 1999.

[79] R. Adler, "A study of locking phenomena in oscillators," *Proceedings of IRE*, vol. 34, pp. 351–357, June 1946.

[80] H. R. Rategh, H. Samavati, and T. H. Lee, "A 5GHz, 1mW CMOS voltage-controlled differential injection locked frequency divider," in *Proceedings of Custom Integrated Circuits Conference*, May 1999, pp. 24.5.1–24.5.4.

[81] H. R. Rategh, H. Samavati, and T. H. Lee, "A 5GHz, 32mW CMOS frequency synthesizer with an injection locked frequency divider," in *Symposium of VLSI Circuits Digest of Technical Papers*, June 1999, pp. 113–116.

[82] H. R. Rategh, H. Samavati, and T. H. Lee, "A CMOS frequency synthesizer with an injection-locked frequency divider for a 5-GHz wireless LAN receiver," *IEEE Journal of Solid-State Circuits*, vol. 35, no. 5, pp. 780–787, May 2000.

[83] D. Li and Y. Tsividis, "Active LC filters on silicon," *IEE Proceedings: Circuits, Devices and Systems*, vol. 147, no. 1, pp. 49–56, February 2000.

[84] M. A. Copeland, S. P. Voinigescu, D. Marchesan, P. Popescu, and M. C. Maliepaard, "5-GHz SiGe HBT monolithic radio transceiver with tunable filtering," *IEEE Trans. on Microwave Theory and Techniques*, vol. 48, no. 2, pp. 170–181, February 2000.

[85] S. P. Voinigescu, M. A. Copeland, D. Marchesan, P. Popescu, and M. C. Maliepaard, "5-GHz SiGe HBT monolithic radio transceiver with tunable filtering," in *Proceedings of IEEE Radio Frequency Integrated Circuits Symposium*, 1999, pp. 131–134.

[86] J. Macedo, M. A. Copeland, and P. Schvan, "A 2.5GHz monolithic silicon image reject filter," in *Proceedings of Custom Integrated Circuits Conference*, May 1996, pp. 10.3.1–10.3.4.

[87] S. Pipilos and Y. Tsividis, "RLC active filters with electronically tunable center frequency and quality factor," *Electronics Letters*, vol. 30, no. 6, pp. 472–474, March 1994.

[88] S. Pipilos and Y. Tsividis, "Design of active RLC integrated filters with application in the GHz range," in *Proceedings of IEEE ISCAS*, 1994, pp. 645–648.

[89] R. A. Duncan, K. W. Martin, and A. S. Sedra, "A Q-enhanced active-RLC bandpass filter," in *Proceedings of IEEE ISCAS*, 1993, pp. 1416–1419.

[90] R. A. Duncan, K. W. Martin, and A. S. Sedra, "A Q-enhanced active-RLC bandpass filter," *IEEE Trans. on Circuits and Systems II*, vol. 44, no. 5, pp. 341–347, May 1997.

[91] S. Pipilos, Y. Tsividis, and J. Fenk, "1.8GHz tunable filter in si technology," in

Proceedings of Custom Integrated Circuits Conference, May 1996, pp. 10.2.1–10.2.4.

[92] W. Gao and W. M. Snelgrove, "A linear active Q-enhanced monolithic LC filter," in *Proceedings of IEEE ISCAS*, June 1997, pp. 97–100.

[93] W. Gao and W. M. Snelgrove, "A linear integrated LC bandpass filter with Q-enhancement," *IEEE Trans. on Circuits and Systems II*, vol. 45, no. 5, pp. 635–639, May 1998.

[94] W. B. Kuhn, N. K. Yanduru, and A. S. Wyszynski, "A high dynamic range, digitally tuned, Q-enhanced LC bandpass filter for cellular/PCS receivers," in *Proceedings of IEEE Radio Frequency Integrated Circuits Symposium*, 1998, pp. 261–264.

[95] W. S. T. Yan, R. K. C. Mak, and H. C. Luong, "2-V 0.8-μm CMOS monolithic RF filter for GSM receivers," in *IEEE Microwave Theory and Techniques Symposium Digest of Technical Papers*, 1999, pp. 569–572.

[96] T. Soorapanth and S. S. Wong, "A 0dB-IL,2140$\pm$30 MHz bandpass filter utilizing Q-enhanced spiral inductors in standard CMOS," in *Symposium of VLSI Circuits Digest of Technical Papers*, June 2001, pp. 15–18.

[97] A. N. Mohieldin, E. Sánchez-Sinencio, and J. S. Martínez, "A 2.7V 1.8GHz fourth-order tunable LC bandpass filter based on emulation of magnetically coupled resonators," *IEEE Journal of Solid-State Circuits*, vol. 38, no. 7, pp. 1172–1181, July 2003.

[98] H. Ahmed, C. DeVries, and R. Mason, "RF, Q-enhanced bandpass filters in standard 0.18μm CMOS with direct digital tuning," in *Proceedings of the*

IEEE International Symposium on Circuits and Systems (ISCAS), Bangkok, Thailand, May 2003, pp. I–577 – I–580.

[99] W. B. Kuhn, D. Nobbe, D. Kelly, and A. W. Orsborn, "Dynamic range performance of on-chip RF bandpass filters," *IEEE Trans. on Circuits and Systems II*, vol. 50, no. 10, pp. 685–694, October 2003.

[100] Y. Wu, X. Ding, M. Ismail, and H. Olsson, "CMOS active inductor and its application in RF bandpass filter," in *Digest of papers, IEEE Radio Frequency Integrated Circuits (RFIC) Symposium*, Fort Worth, June 2004, pp. 655–658.

[101] W. A. Gee and P. E. Allen, "CMOS integrated LC RF bandpass filter with transformer-coupled Q-enhancement and optimized linearity," in *Proceedings of the IEEE International Symposium on Circuits and Systems (ISCAS)*, New Orleans, May 2007, pp. 1445–1448.

[102] D. Shi and M. P. Flynn, "A compact 5GHz Q-enhanced standing-wave resonator based filter in 0.13μm CMOS," in *Digest of papers, IEEE Radio Frequency Integrated Circuits (RFIC) Symposium*, Atlanta, June 2008, pp. 453–456.

[103] Z. Gao, J. Ma, M. Yu, and Y. Ye, "A fully integrated CMOS active bandpass filter for multiband RF front-ends," *IEEE Trans. on Circuits and Systems II*, vol. 55, no. 8, pp. 718–722, August 2008.

[104] P. Wambacq and W. Sansen, *Distortion Analysis of Analog Integrated Circuits*, Kluwer Academic Publishers, Boston, 1998.

[105] Y. P. Tsividis, "Integrated continuous-time filter design-an overview," *IEEE Journal of Solid-State Circuits*, vol. 29, no. 3, March 1994.

[106] M. R. Spiegel and J. Liu, *Mathematical Handbook of Formulas and Tables*, McGraw Hill, New York, 2nd edition, 1999.

[107] D. Y. Aksın, "11-bits sub-ranging analog to digital converter and SSAtool," Ph.D. dissertation, The University of Texas at Dallas, Dallas, 2006.

[108] H. Samavati, H. R. Rategh, and T. H. Lee, "A 5-GHz CMOS wireless LAN receiver front end," *IEEE Journal of Solid-State Circuits*, vol. 35, no. 5, pp. 765–772, May 2000.

[109] B. Razavi, *Design of Analog CMOS Integrated Circuits*, McGraw-Hill, New York, NY, 2001.

[110] G. Gonzalez, *Microwave Transistor Amplifiers, Analysis and Design*, Prentice Hall, Upper Saddle River, NJ, 2nd edition, 1997.

[111] A. Zanchi, C. Samori, S. Levantino, and A. L. Lacaita, "A 2V 2.5GHz -104dBc/Hz at 100kHz fully integrated VCO with wide-band low-noise automatic amplitude control loop," *IEEE Journal of Solid-State Circuits*, vol. 36, no. 4, pp. 611–619, April 2001.

[112] M. Zannoth, B. Kolb, J. Fenk, and R. Weigel, "A fully integrated VCO at 2GHz," *IEEE Journal of Solid-State Circuits*, vol. 33, no. 12, pp. 1987–1991, December 1998.

[113] J. O. Plouchart, H. Ainspan, M. Soyuer, and A. Ruehli, "A fully monolithic SiGe differential VCO for 5GHz wireless applications," in *Proceedings of IEEE RF IC Symposium*, Boston, June 2000, pp. 57–60.

[114] E. Hegazi, H. Sjöland, and A. Abidi, "A filtering technique to lower oscillator phase noise," in *International Solid-State Circuits Conference Digest of*

Technical Papers, San Francisco, February 2001, pp. 364–365.

[115] H. Wang, "A 50GHz VCO in 0.25μm CMOS," in *International Solid-State Circuits Conference Digest of Technical Papers*, San Francisco, February 2001, pp. 372–373.

[116] M. Tiebout, "A differentially tuned 1.73GHz-1.99GHz quadrature CMOS VCO for DECT, DCS1800 and GSM900 with a phase noise over tuning range between -128dBc/Hz and -137dBc/Hz at 600kHz offset," in *Proceedings of European Solid-State Circuits Conference*, September 2000, pp. 444–447.

APPENDIX A

DIRECT CALCULATION OF NONLINEAR RESPONSES

The derivation of the 1dB compression point of the filter with only the nonlinearity of the input G_m stage taken into account is presented in this appendix. As an illustration of the direct calculation of nonlinear responses method proposed in [104], the nonlinearity of a differential pair with and without source-degeneration resistors are analyzed. Recall that the input G_m stage is a differential pair with source degeneration resistors.

1. Nonlinearity of a MOS differential pair

The circuit schematic of a differential pair without source-degeneration resistors is shown in Fig.155. The load impedances are denoted as Z_L and in the case of the filter Z_L is the parallel equivalent of the lossy LC tank and the negative conductance generated by the cross-coupled pair.

The AC equivalent circuit of the differential pair is depicted in Fig.156. As the two input transistors are assumed to be identical, their transconductances and the corresponding higher-order nonlinearity coefficients will be represented by the same symbol in the following derivations. The output conductances and the bulk-source transconductances of the transistors are neglected for the sake of simplicity. It should be noted that only the nonlinearity of the transconductance of the devices are taken into account.

The direct calculation of nonlinear responses method consists of three steps, [104]. In the first step, the computation of the first-order (fundamental) responses

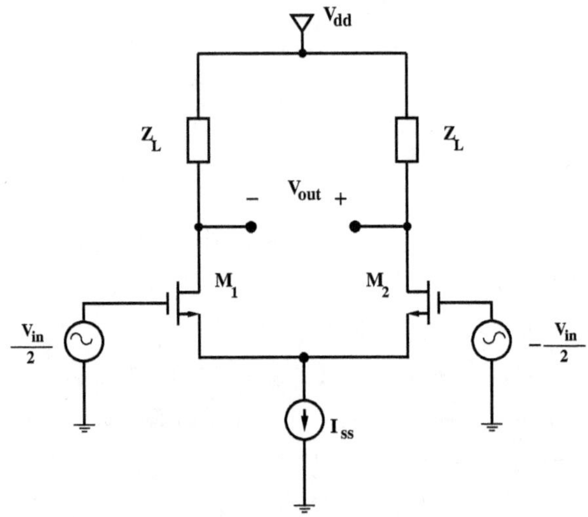

Fig. 155. Differential pair with differential excitation at its inputs.

are carried out using the linearized network given in Fig.157. The equations for the node voltages can be written from Fig.157 as

$$\begin{aligned} g_m\left(V_{4,1}-V_{3,1}\right)+g_m\left(V_{5,1}-V_{3,1}\right) &= 0 \\ V_{1,1}\cdot Y_L+g_m\left(V_{4,1}-V_{3,1}\right) &= 0 \\ V_{2,1}\cdot Y_L+g_m\left(V_{5,1}-V_{3,1}\right) &= 0 \end{aligned} \tag{A.1}$$

where $V_{i,1}$ denotes the voltage component with fundamental frequency at node i. The input signals at nodes 4 and 5 are assumed to be pure sinusoidal signals which have components only at the fundamental frequency, ω_1. The input signals are given as $V_{4,1}=-V_{5,1}=V_{in}/2$. Substituting the input voltages into (A.1) and solving for the node voltages at the fundamental frequencies, the following is obtained

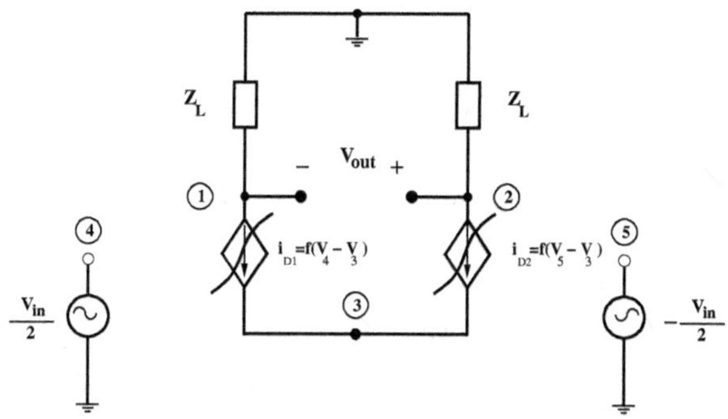

Fig. 156. AC equivalent circuit of the differential pair.

$$
\begin{aligned}
V_{3,1} &= 0 \\
V_{1,1} &= -\frac{g_m}{2Y_L} \cdot V_{in} \\
V_{2,1} &= \frac{g_m}{2Y_L} \cdot V_{in}.
\end{aligned} \tag{A.2}
$$

As expected, the voltage at the common source node does not contain any component at the fundamental frequency, i.e. $V_{3,1} = 0$. In other words, the common source node acts as a virtual ground for the fundamental frequency once the mismatches are neglected due to the fact that the circuit is driven differentially. The output voltage at ω_1 can now be expressed as

$$
V_{out,1} = V_{2,1} - V_{1,1} = \frac{g_m}{Y_L} \cdot V_{in}. \tag{A.3}
$$

Now that the first-order responses are computed, the second step in the method can be carried out. The second step is to calculate the second-order responses. The second-order nonlinearity in the circuit will give rise to responses at 0 (DC) and $2\omega_1$.

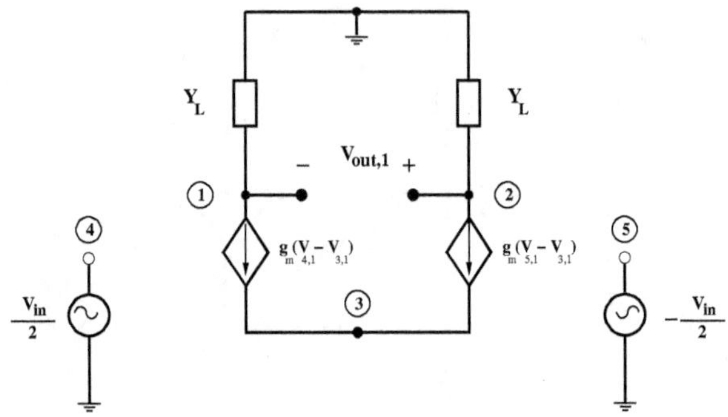

Fig. 157. Linearized equivalent circuit for the calculation of the first-order responses.

In the analysis the same linearized network is to be solved, but this time with the external excitations neutralized and nonlinear current sources of order two connected in parallel to the linearized equivalent of the nonlinearity, as shown in Fig.158. As explained in [104], in the Volterra Series approach, nonlinear current sources are applied in order to calculate the higher-order Volterra kernels. The current sources are determined by the lower-order Volterra kernels of the voltages that control the corresponding nonlinearities. In the direct calculation of the nonlinear responses, a similar approach is proposed in [104]. Namely, the values of the nonlinear current sources depend on the lower-order voltages that control the nonlinearities. The nonlinear current source associated with transconductance for response at $2\omega_1$ is provided in [104] as

$$i_{NL2_{gm}} \equiv \frac{K_{2_{gm}}}{2} \cdot (V_{i,1})^2 \, . \tag{A.4}$$

The controlling voltages of the nonlinear current sources of order two to be computed are $V_{43,1}$ and $V_{53,1}$ given as

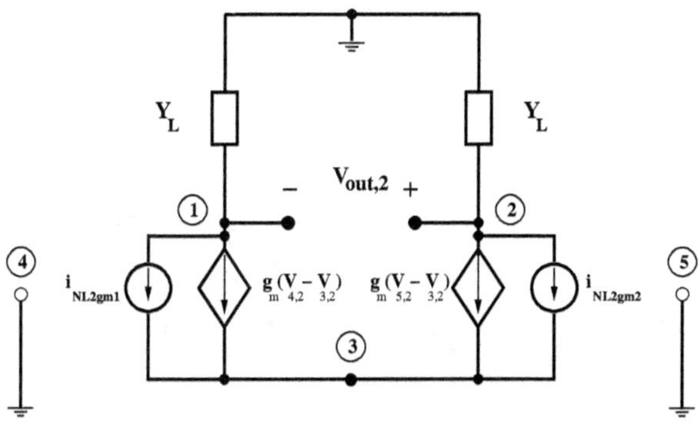

Fig. 158. Linearized equivalent circuit for the calculation of the second-order responses.

$$\begin{aligned} V_{43,1} &= V_{4,1} - V_{3,1} = \frac{V_{in}}{2} \\ V_{53,1} &= V_{5,1} - V_{3,1} = -\frac{V_{in}}{2}. \end{aligned} \tag{A.5}$$

Using (A.4) and (A.5), the values of the nonlinear current sources are computed as

$$\begin{aligned} i_{NL2_{gm1}} &\equiv \frac{K_{2_{gm1}}}{2} \cdot (V_{43,1})^2 = \frac{K_{2_{gm}} \cdot V_{in}^2}{8} \\ i_{NL2_{gm2}} &\equiv \frac{K_{2_{gm2}}}{2} \cdot (V_{53,1})^2 = \frac{K_{2_{gm}} \cdot V_{in}^2}{8}. \end{aligned} \tag{A.6}$$

It is observed in (A.6) that the values of the two nonlinear current sources are equal as the mismatches are neglected. The equations for the node voltages can be written from Fig.158 as

$$i_{NL2_{gm}} + g_m (V_{4,2} - V_{3,2}) + i_{NL2_{gm}} + g_m (V_{5,2} - V_{3,2}) = 0$$

$$V_{1,2} \cdot Y_L + g_m \left(V_{4,2} - V_{3,2}\right) + i_{NL2_{gm}} = 0$$
$$V_{2,2} \cdot Y_L + g_m \left(V_{5,2} - V_{3,2}\right) + i_{NL2_{gm}} = 0 \qquad \text{(A.7)}$$

where $V_{i,2}$ denotes the voltage component with twice the fundamental frequency at node i. As the input signals at nodes 4 and 5 are assumed to be pure sinusoidal signals which have components only at the fundamental frequency, $V_{4,2} = V_{5,2} = 0$. Substituting the input voltages and (A.6) into (A.7) and then solving for the node voltage second-order responses yield

$$\begin{aligned} V_{3,2} &= \frac{K_{2_{gm}} \cdot V_{in}^2}{8 g_m} \\ V_{1,2} &= \left(-i_{NL2_{gm}} + g_m V_{3,2}\right) \frac{1}{Y_L} \\ V_{2,2} &= \left(-i_{NL2_{gm}} + g_m V_{3,2}\right) \frac{1}{Y_L}. \end{aligned} \qquad \text{(A.8)}$$

It is important to point out that the voltage at the common source node does not act as a virtual ground for twice the fundamental frequency, i.e. $V_{3,2} \neq 0$. Recall that the second-order nonlinearity also produces components at DC. Namely, a DC component the value of which depends on the input signal amplitude exists on the common-source node due to the second-order nonlinearity. On the other hand, the output voltage at $2\omega_1$ is given as

$$V_{out,2} = V_{2,2} - V_{1,2} = 0 \qquad \text{(A.9)}$$

since the mismatches are neglected.

Now that the second-order responses are computed, the third step in the method can be carried out. The third step is to calculate the third-order responses. The third-order nonlinearity in the circuit will give rise to responses at ω_1 and $3\omega_1$. Once

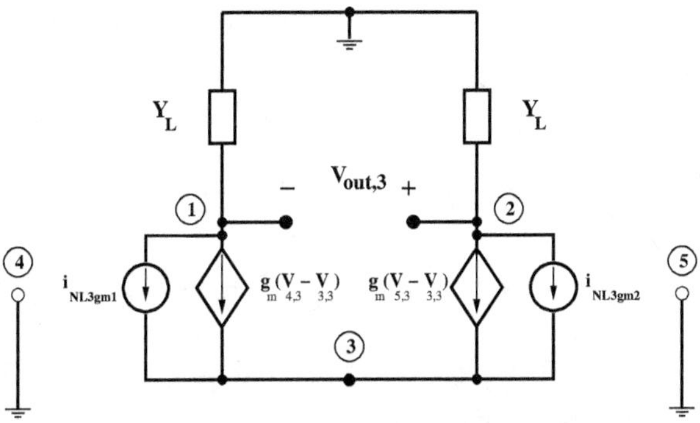

Fig. 159. Linearized equivalent circuit for the calculation of the third-order responses.

again, in the analysis the same linearized network as in the case of second-order response analysis, is to be solved, but this time with the nonlinear current sources of order three connected in parallel to the linearized equivalent of the nonlinearity, as shown in Fig.159. The nonlinear current source associated with transconductance for response at $3\omega_1$ is provided in [104] as

$$i_{NL3_{gm}} \equiv K_{2_{gm}} V_{i,1} V_{i,2} + \frac{1}{4} K_{3_{gm}} V_{i,1}^3. \tag{A.10}$$

The controlling voltages of the nonlinear current sources of order three to be computed are $V_{43,1}$, $V_{53,1}$, $V_{43,2}$ and $V_{53,2}$. The controlling voltages of order one are given in (A.5) and those of order two can be computed using the second-order responses found above as following

$$\begin{aligned} V_{43,2} &= V_{4,2} - V_{3,2} = -\frac{K_{2_{gm}} V_{in}^2}{8 g_m} \\ V_{53,2} &= V_{5,2} - V_{3,2} = -\frac{K_{2_{gm}} V_{in}^2}{8 g_m}. \end{aligned} \tag{A.11}$$

Using (A.5), (A.10) and (A.11), the values of the nonlinear current sources of order three are computed as

$$\begin{aligned} i_{NL3_{gm1}} &= -\frac{K_{2_{gm}}^2 V_{in}^3}{16 g_m} + \frac{1}{32} K_{3_{gm}} V_{in}^3 \\ i_{NL3_{gm2}} &= \frac{K_{2_{gm}}^2 V_{in}^3}{16 g_m} - \frac{1}{32} K_{3_{gm}} V_{in}^3. \end{aligned} \tag{A.12}$$

It is observed in (A.12) that $i_{NL3_{gm1}} = -i_{NL3_{gm2}}$. The equations for the node voltages can be written from Fig.159 as

$$\begin{aligned} i_{NL3_{gm1}} + g_m (V_{4,3} - V_{3,3}) + i_{NL3_{gm2}} + g_m (V_{5,3} - V_{3,3}) &= 0 \\ V_{1,3} \cdot Y_L + g_m (V_{4,3} - V_{3,3}) + i_{NL3_{gm1}} &= 0 \\ V_{2,3} \cdot Y_L + g_m (V_{5,3} - V_{3,3}) + i_{NL3_{gm2}} &= 0 \end{aligned} \tag{A.13}$$

where $V_{i,3}$ denotes the voltage component with three times the fundamental frequency at node i. As the input signals at nodes 4 and 5 are assumed to be pure sinusoidal signals which have components only at the fundamental frequency, $V_{4,3} = V_{5,3} = 0$. Substituting the input voltages and (A.12) into (A.13) and then solving for the node voltage third-order responses yield

$$\begin{aligned} V_{3,3} &= 0 \\ V_{1,3} &= \frac{1}{Y_L} \left(\frac{K_{2_{gm}}^2 V_{in}^3}{16 g_m} - \frac{1}{32} K_{3_{gm}} V_{in}^3 \right) \\ V_{2,3} &= -\frac{1}{Y_L} \left(\frac{K_{2_{gm}}^2 V_{in}^3}{16 g_m} - \frac{1}{32} K_{3_{gm}} V_{in}^3 \right). \end{aligned} \tag{A.14}$$

It is seen that the voltage at the common source node acts as a virtual ground for also three times the fundamental frequency. In other words, the common-source node is virtual ground for odd order harmonics of the input frequency including the fundamental frequency itself, with the assumption of perfect matching. On the other hand, the output voltage due to the third-order nonlinearities can be obtained from (A.14) as

$$V_{out,3} = V_{2,3} - V_{1,3} = -\frac{2}{Y_L}\left(\frac{K_{2_{gm}}^2}{16g_m} - \frac{K_{3_{gm}}}{32}\right)V_{in}^3. \tag{A.15}$$

As mentioned before and also elaborated in the nonlinearity analyses presented in Chapter IV, the output due to the third-order nonlinearity has a nonlinear component at both the fundamental frequency and the third-order harmonic of the input frequency. Let $V_{in} = V\cos(\omega_1 t)$ for a sinusoidal input. With the trigonometric relationship of $\cos^3(\omega_1 t) = \frac{1}{4}(3\cos(\omega_1 t) + \cos(3\omega_1 t))$, the following expression is obtained for the output voltage at the fundamental frequency using (A.3) and (A.15)

$$V_{out} = \left(\overbrace{\left(\frac{g_m}{Y_L(j\omega_1)}\right)}^{\textit{linear term}} - \overbrace{\frac{3}{4}\frac{2}{Y_L(j\omega_1)}\left(\frac{K_{2_{gm}}^2}{16g_m} - \frac{K_{3_{gm}}}{32}\right)V^2}^{\textit{nonlinear term}}\right)V\cos(\omega_1 t). \tag{A.16}$$

Note that the nonlinear term causes compression at the output as the input amplitude increases (it should be recalled that the third-order nonlinearity coefficient of the transconductance gain of a transistor, $K_{3_{gm}}$, is negative as given in (4.63)). Carrying out the required algebra to calculate the input amplitude where the output is compressed by 1dB yields the following expression for the input 1dB compression point

$$V_{in_{1dB}} \simeq \sqrt{\frac{2.32\, g_m^2}{2K_{2_{gm}}^2 - K_{3_{gm}} g_m}}. \tag{A.17}$$

2. Nonlinearity of a MOS differential pair with source-degeneration

The circuit schematic of a differential pair with source-degeneration resistors is shown in Fig.160. The load impedances are denoted as Z_L and in the case of the filter Z_L is the parallel equivalent of the lossy LC tank and the negative conductance generated by the cross-coupled pair.

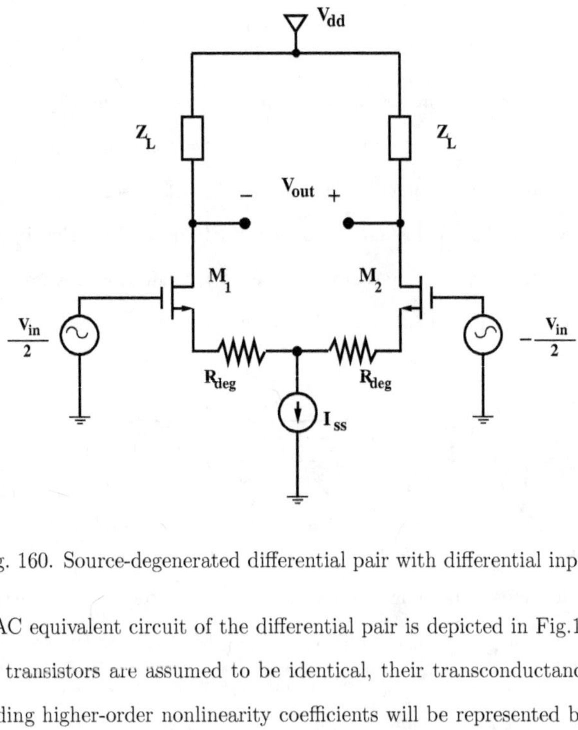

Fig. 160. Source-degenerated differential pair with differential inputs.

The AC equivalent circuit of the differential pair is depicted in Fig.161. As the two input transistors are assumed to be identical, their transconductances and the corresponding higher-order nonlinearity coefficients will be represented by the same symbol in the following derivations. The output conductances and the bulk-source transconductances of the transistors are neglected for the sake of simplicity. It should be noted that only the nonlinearity of the transconductance of the devices are taken into account.

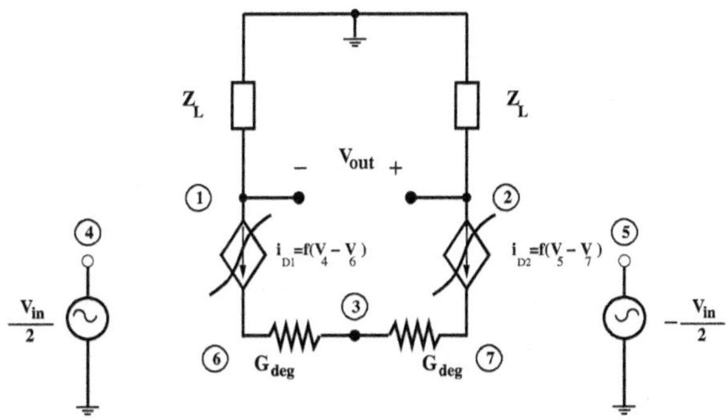

Fig. 161. AC equivalent circuit of the differential pair with source-degeneration.

In the first step, the computation of the first-order (fundamental) responses are carried out using the linearized network given in Fig.162. The equations for the node voltages can be written from Fig.162 as

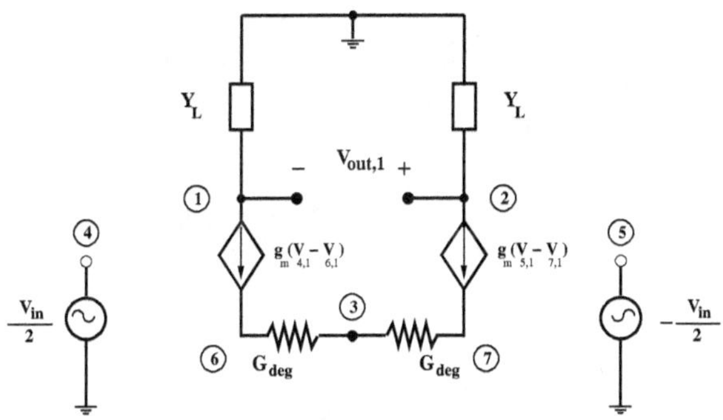

Fig. 162. Linearized equivalent circuit with source-degeneration for the calculation of the first-order responses.

$$g_m \left(V_{4,1} - V_{6,1}\right) - G_{deg}\left(V_{6,1} - V_{3,1}\right) \;=\; 0$$

$$
\begin{aligned}
V_{1,1} \cdot Y_L + g_m \left(V_{4,1} - V_{6,1}\right) &= 0 \\
G_{deg} \left(V_{6,1} - V_{3,1}\right) + G_{deg} \left(V_{7,1} - V_{3,1}\right) &= 0 \\
V_{2,1} \cdot Y_L + g_m \left(V_{5,1} - V_{7,1}\right) &= 0 \\
g_m \left(V_{5,1} - V_{7,1}\right) - G_{deg} \left(V_{7,1} - V_{3,1}\right) &= 0
\end{aligned} \tag{A.18}
$$

where $V_{i,1}$ denotes the voltage component with fundamental frequency at node i. The input signals at nodes 4 and 5 are assumed to be pure sinusoidal signals which have components only at the fundamental frequency, ω_1. The input signals are given as $V_{4,1} = -V_{5,1} = V_{in}/2$. Substituting the input voltages into (A.18) and solving for the node voltages at the fundamental frequencies, the following is obtained

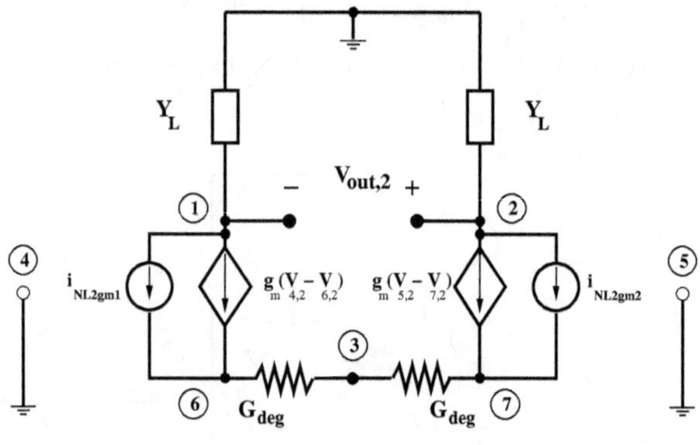

Fig. 163. Linearized equivalent circuit with source-degeneration for the calculation of the second-order responses.

$$
\begin{aligned}
V_{3,1} &= 0 \\
V_{6,1} &= \frac{g_m R_{deg}}{1 + g_m R_{deg}} \cdot \frac{v_{in}}{2} \\
V_{7,1} &= -\frac{g_m R_{deg}}{1 + g_m R_{deg}} \cdot \frac{v_{in}}{2}
\end{aligned}
$$

$$\begin{aligned} V_{1,1} &= -\frac{1}{Y_L}\frac{g_m}{1+g_m R_{deg}}\cdot\frac{v_{in}}{2} \\ V_{2,1} &= \frac{1}{Y_L}\frac{g_m}{1+g_m R_{deg}}\cdot\frac{v_{in}}{2}. \end{aligned} \tag{A.19}$$

Note that the voltage at the common mode node where the tail current source is connected does not contain any component at the fundamental frequency, i.e. $V_{3,1} = 0$. In other words, the common mode node acts as a virtual ground for the fundamental frequency once the mismatches are neglected due to the fact that the circuit is driven differentially. The output voltage at ω_1 can now be expressed as

$$V_{out,1} = V_{2,1} - V_{1,1} = \frac{1}{Y_L}\frac{g_m}{1+g_m R_{deg}} V_{in}. \tag{A.20}$$

Now that the first-order responses are computed, the second step in the method can be carried out. The second step is to calculate the second-order responses. The second-order nonlinearity in the circuit will give rise to responses at 0 (DC) and $2\omega_1$. In the analysis the same linearized network is to be solved, but this time with the external excitations neutralized and nonlinear current sources of order two connected in parallel to the linearized equivalent of the nonlinearity, as shown in Fig.163. Recall that, the values of the nonlinear current sources depend on the lower-order voltages that control the nonlinearities. The nonlinear current source associated with transconductance for response at $2\omega_1$ is provided in [104] as

$$i_{NL2_{gm}} \equiv \frac{K_{2_{gm}}}{2} \cdot (V_{i,1})^2. \tag{A.21}$$

The controlling voltages of the nonlinear current sources of order two to be computed are $V_{46,1}$ and $V_{57,1}$ given as

$$
\begin{aligned}
V_{46,1} &= V_{4,1} - V_{6,1} = \frac{V_{in}}{2}\frac{1}{1+g_m R_{deg}} \\
V_{57,1} &= V_{5,1} - V_{7,1} = -\frac{V_{in}}{2}\frac{1}{1+g_m R_{deg}}.
\end{aligned} \tag{A.22}
$$

Using (A.21) and (A.22), the values of the nonlinear current sources are computed as

$$
\begin{aligned}
i_{NL2_{gm1}} &\equiv \frac{K_{2_{gm1}}}{2}\cdot (V_{46,1})^2 = \frac{K_{2_{gm}}\cdot V_{in}^2}{8\,(1+g_m R_{deg})^2} \\
i_{NL2_{gm2}} &\equiv \frac{K_{2_{gm2}}}{2}\cdot (V_{57,1})^2 = \frac{K_{2_{gm}}\cdot V_{in}^2}{8\,(1+g_m R_{deg})^2}.
\end{aligned} \tag{A.23}
$$

It is observed in (A.23) that the values of the two nonlinear current sources are equal as the mismatches are neglected. The equations for the node voltages can be written from Fig.163 as

$$
\begin{aligned}
i_{NL2_{gm}} + g_m\,(V_{4,2} - V_{6,2}) - G_{deg}\,(V_{6,2} - V_{3,2}) &= 0 \\
V_{1,2}\cdot Y_L + g_m\,(V_{4,2} - V_{6,2}) + i_{NL2_{gm}} &= 0 \\
G_{deg}\,(V_{6,2} - V_{3,2}) + G_{deg}\,(V_{7,2} - V_{3,2}) &= 0 \\
V_{2,2}\cdot Y_L + g_m\,(V_{5,2} - V_{7,2}) + i_{NL2_{gm}} &= 0 \\
i_{NL2_{gm}} + g_m\,(V_{5,2} - V_{7,2}) - G_{deg}\,(V_{7,2} - V_{3,2}) &= 0
\end{aligned} \tag{A.24}
$$

where $V_{i,2}$ denotes the voltage component with twice the fundamental frequency at node i. As the input signals at nodes 4 and 5 are assumed to be pure sinusoidal signals which have components only at the fundamental frequency, $V_{4,2} = V_{5,2} = 0$. Substituting the input voltages and (A.23) into (A.24) and then solving for the node voltage second-order responses, yield

$$\begin{aligned}
V_{3,2} &= \frac{i_{NL2_{gm}}}{g_m} \\
V_{6,2} &= \frac{i_{NL2_{gm}}}{g_m} \\
V_{7,2} &= \frac{i_{NL2_{gm}}}{g_m} \\
V_{1,2} &= 0 \\
V_{2,2} &= 0.
\end{aligned} \tag{A.25}$$

It is important to point out that the voltage at the common mode node does not act as a virtual ground for twice the fundamental frequency, i.e. $V_{3,2} \neq 0$. Recall that the second-order nonlinearity also produces components at DC. Namely, a DC component the value of which depends on the input signal amplitude exists on the common mode node due to the second-order nonlinearity. On the other hand, the output voltage at $2\omega_1$ is given as

$$V_{out,2} = V_{2,2} - V_{1,2} = 0 \tag{A.26}$$

since the mismatches are neglected.

Now that the second-order responses are computed, the third step in the method can be carried out. The third step is to calculate the third-order responses. The third-order nonlinearity in the circuit will give rise to responses at ω_1 and $3\omega_1$. Once again, in the analysis the same linearized network as in the case of second-order response analysis, is to be solved, but this time with the nonlinear current sources of order three connected in parallel to the linearized equivalent of the nonlinearity, as shown in Fig.164. The nonlinear current source associated with transconductance for response at $3\omega_1$ is provided in [104] as

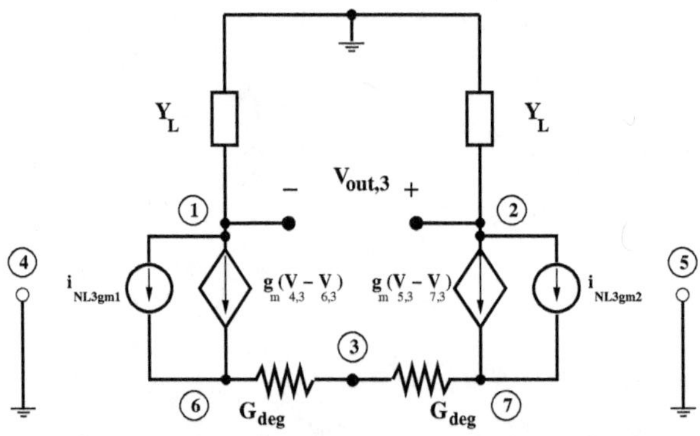

Fig. 164. Linearized equivalent circuit with source-degeneration for the calculation of the third-order responses.

$$i_{NL3_{gm}} \equiv K_{2_{gm}} V_{i,1} V_{i,2} + \frac{1}{4} K_{3_{gm}} V_{i,1}^3. \tag{A.27}$$

The controlling voltages of the nonlinear current sources of order three to be computed are $V_{46,1}$, $V_{57,1}$, $V_{46,2}$ and $V_{57,2}$. The controlling voltages of order one are given in (A.22) and those of order two can be computed using the second-order responses found above as following

$$\begin{aligned} V_{46,2} &= V_{4,2} - V_{6,2} = -\frac{K_{2_{gm}} V_{in}^2}{8 g_m \left(1 + g_m R_{deg}\right)^2} \\ V_{57,2} &= V_{5,2} - V_{7,2} = -\frac{K_{2_{gm}} V_{in}^2}{8 g_m \left(1 + g_m R_{deg}\right)^2}. \end{aligned} \tag{A.28}$$

Using (A.22), (A.27) and (A.28), the values of the nonlinear current sources of order three are computed as

$$\begin{aligned} i_{NL3_{gm1}} &= -\frac{K_{2_{gm}}^2 V_{in}^3}{16 g_m \left(1 + g_m R_{deg}\right)^3} + \frac{K_{3_{gm}} V_{in}^3}{32 \left(1 + g_m R_{deg}\right)^3} \\ i_{NL3_{gm2}} &= \frac{K_{2_{gm}}^2 V_{in}^3}{16 g_m \left(1 + g_m R_{deg}\right)^3} - \frac{K_{3_{gm}} V_{in}^3}{32 \left(1 + g_m R_{deg}\right)^3}. \end{aligned} \quad (A.29)$$

It is observed in (A.29) that $i_{NL3_{gm1}} = -i_{NL3_{gm2}}$. It should also be pointed out that a comparison between (A.12) and (A.29) shows a reduction in the values of the nonlinear current sources of order three by a factor of $(1 + g_m R_{deg})^3$ with the source-degeneration resistors. The equations for the node voltages can be written from Fig.164 as

$$\begin{aligned} i_{NL3_{gm1}} + g_m \left(V_{4,3} - V_{6,3}\right) - G_{deg} \left(V_{6,3} - V_{3,3}\right) &= 0 \\ V_{1,3} \cdot Y_L + g_m \left(V_{4,3} - V_{6,3}\right) + i_{NL3_{gm1}} &= 0 \\ G_{deg} \left(V_{6,3} - V_{3,3}\right) + G_{deg} \left(V_{7,3} - V_{3,3}\right) &= 0 \\ V_{2,3} \cdot Y_L + g_m \left(V_{5,3} - V_{7,3}\right) + i_{NL3_{gm2}} &= 0 \\ i_{NL3_{gm2}} + g_m \left(V_{5,3} - V_{7,3}\right) - G_{deg} \left(V_{7,3} - V_{3,3}\right) &= 0 \end{aligned} \quad (A.30)$$

where $V_{i,3}$ denotes the voltage component with three times the fundamental frequency at node i. As the input signals at nodes 4 and 5 are assumed to be pure sinusoidal signals which have components only at the fundamental frequency, $V_{4,3} = V_{5,3} = 0$. Substituting the input voltages and (A.29) into (A.30) and then solving for the node voltage third-order responses yield

$$\begin{aligned} V_{3,3} &= 0 \\ V_{6,3} &= -\frac{R_{deg} K_{2_{gm}}^2 V_{in}^3}{16 g_m \left(1 + g_m R_{deg}\right)^4} + \frac{R_{deg} K_{3_{gm}} V_{in}^3}{32 \left(1 + g_m R_{deg}\right)^4} \end{aligned}$$

$$
\begin{aligned}
V_{7,3} &= \frac{R_{deg}K_{2_{gm}}^2 V_{in}^3}{16g_m\left(1+g_mR_{deg}\right)^4} - \frac{R_{deg}K_{3_{gm}} V_{in}^3}{32\left(1+g_mR_{deg}\right)^4} \\
V_{1,3} &= \frac{1}{Y_L}\left(\frac{K_{2_{gm}}^2 V_{in}^3}{16g_m\left(1+g_mR_{deg}\right)^4} - \frac{K_{3_{gm}} V_{in}^3}{32\left(1+g_mR_{deg}\right)^4}\right) \\
V_{2,3} &= \frac{1}{Y_L}\left(-\frac{K_{2_{gm}}^2 V_{in}^3}{16g_m\left(1+g_mR_{deg}\right)^4} + \frac{K_{3_{gm}} V_{in}^3}{32\left(1+g_mR_{deg}\right)^4}\right).
\end{aligned} \tag{A.31}
$$

It is seen that the voltage at the common mode node acts as a virtual ground for also three times the fundamental frequency. In other words, the common mode node is virtual ground for odd order harmonics of the input frequency including the fundamental frequency itself, with the assumption of perfect matching. On the other hand, the output voltage due to the third-order nonlinearities can be obtained from (A.31) as

$$
V_{out,3} = V_{2,3} - V_{1,3} = -\frac{2}{Y_L}\left(\frac{K_{2_{gm}}^2}{16g_m} - \frac{K_{3_{gm}}}{32}\right)\frac{V_{in}^3}{\left(1+g_mR_{deg}\right)^4}. \tag{A.32}
$$

A comparison between (A.15) and (A.32) shows a reduction in the value of the output voltage due to the third-order nonlinearities in the circuit by a factor of $(1+g_mR_{deg})^4$ with the source-degeneration resistors. As mentioned before and also elaborated in the nonlinearity analyses presented in Chapter IV, the output due to the third-order nonlinearity has a nonlinear component at both the fundamental frequency and the third-order harmonic of the input frequency. Let $V_{in} = V\cos(\omega_1 t)$ for a sinusoidal input. With the trigonometric relationship of $\cos^3(\omega_1 t) = \frac{1}{4}(3\cos(\omega_1 t) + \cos(3\omega_1 t))$, the following expression is obtained for the output voltage at the fundamental frequency using (A.20) and (A.32)

$$V_{out} = \left(\overbrace{\left(\frac{g_m}{B\ Y_L(j\omega_1)} \right)}^{linear\ term} - \overbrace{\frac{3}{4} \frac{2}{B^4\ Y_L(j\omega_1)} \left(\frac{K_{2_{gm}}^2}{16 g_m} - \frac{K_{3_{gm}}}{32} \right) V^2}^{nonlinear\ term} \right) V \cos(\omega_1 t) \quad \text{(A.33)}$$

where $B = (1 + g_m R_{deg})$. Note that the nonlinear term causes compression at the output as the input amplitude increases (it should be recalled that the third-order nonlinearity coefficient of the transconductance gain of a transistor, $K_{3_{gm}}$, is negative as given in (4.63)). Carrying out the required algebra to calculate the input amplitude where the output is compressed by 1dB yields the following expression for the input 1dB compression point

$$V_{in_{1dB}} \simeq \sqrt{\frac{2.32\ g_m^2\ (1 + g_m R_{deg})^3}{2K_{2_{gm}}^2 - K_{3_{gm}} g_m}}. \quad \text{(A.34)}$$

It is important to point out that the input 1dB compression point of a differential pair with source-degeneration is a factor of $\sqrt{(1 + g_m R_{deg})^3}$ higher than that of a simple differential pair as a comparison between (A.17) and (A.34) shows.

www.ingramcontent.com/pod-product-compliance
Lightning Source LLC
LaVergne TN
LVHW022304060326
832902LV00020B/3260
* 9 7 8 3 6 3 9 0 7 4 3 0 7 *